Strengthening Data Science Methods FOR DEPARTMENT OF DEFENSE PERSONNEL AND READINESS MISSIONS

Committee on Strengthening Data Science Methods for Department of Defense Personnel and Readiness Missions

Committee on Applied and Theoretical Statistics

Board on Mathematical Sciences and Their Applications

Division on Engineering and Physical Sciences

A Report of

The National Academies of
SCIENCES · ENGINEERING · MEDICINE

THE NATIONAL ACADEMIES PRESS
Washington, DC
www.nap.edu

THE NATIONAL ACADEMIES PRESS   500 Fifth Street, NW   Washington, DC 20001

This project was supported by Contract No. HQ0034-10-D-0003 TO#12 with the U.S. Department of Defense and Grant No. H98230-14-1-0116 from the National Security Agency. Any opinions, findings, conclusions, or recommendations expressed in this publication do not necessarily reflect the view of any organization or agency that provided support for the project.

International Standard Book Number 13: 978-0-309-45078-2
International Standard Book Number 10: 0-309-45078-0
Digital Object Identifier: 10.17226/23670

Additional copies of this report are available for sale from the National Academies Press, 500 Fifth Street, NW, Keck 360, Washington, DC 20001; (800) 624-6242 or (202) 334-3313; http://www.nap.edu.

Printed in the United States of America

Suggested citation: National Academies of Sciences, Engineering, and Medicine. 2017. *Strengthening Data Science Methods for Department of Defense Personnel and Readiness Missions*. Washington, DC: The National Academies Press. doi:10.17226/23670.

The National Academies of
SCIENCES · ENGINEERING · MEDICINE

The **National Academy of Sciences** was established in 1863 by an Act of Congress, signed by President Lincoln, as a private, nongovernmental institution to advise the nation on issues related to science and technology. Members are elected by their peers for outstanding contributions to research. Dr. Marcia McNutt is president.

The **National Academy of Engineering** was established in 1964 under the charter of the National Academy of Sciences to bring the practices of engineering to advising the nation. Members are elected by their peers for extraordinary contributions to engineering. Dr. C. D. Mote, Jr., is president.

The **National Academy of Medicine** (formerly the Institute of Medicine) was established in 1970 under the charter of the National Academy of Sciences to advise the nation on medical and health issues. Members are elected by their peers for distinguished contributions to medicine and health. Dr. Victor J. Dzau is president.

The three Academies work together as the **National Academies of Sciences, Engineering, and Medicine** to provide independent, objective analysis and advice to the nation and conduct other activities to solve complex problems and inform public policy decisions. The National Academies also encourage education and research, recognize outstanding contributions to knowledge, and increase public understanding in matters of science, engineering, and medicine.

Learn more about the National Academies of Sciences, Engineering, and Medicine at **www.national-academies.org.**

The National Academies of

SCIENCES • ENGINEERING • MEDICINE

Reports document the evidence-based consensus of an authoring committee of experts. Reports typically include findings, conclusions, and recommendations based on information gathered by the committee and committee deliberations. Reports are peer reviewed and are approved by the National Academies of Sciences, Engineering, and Medicine.

Proceedings chronicle the presentations and discussions at a workshop, symposium, or other convening event. The statements and opinions contained in proceedings are those of the participants and have not been endorsed by other participants, the planning committee, or the National Academies of Sciences, Engineering, and Medicine.

For information about other products and activities of the National Academies, please visit nationalacademies.org/whatwedo.

COMMITTEE ON STRENGTHENING DATA SCIENCE METHODS FOR DEPARTMENT OF DEFENSE PERSONNEL AND READINESS MISSIONS

STEPHEN M. ROBINSON, NAE,[1] University of Wisconsin, Madison, *Chair*
PASCALE CARAYON, University of Wisconsin, Madison
DAVID CHU, Institute for Defense Analyses, Alexandria, Virginia
CYNTHIA DWORK, NAS[2]/NAE, Microsoft Research, Mountain View, California
TERRY P. HARRISON, Pennsylvania State University, University Park
ALAN F. KARR, RTI International, Research Triangle Park, North Carolina
SALLIE KELLER, Social Decision Analytics Laboratory, Biocomplexity Institute of Virginia Tech, Arlington
ALAIR MACLEAN, Washington State University, Vancouver
DAVID MAIER, Portland State University, Oregon
STEPHEN M. POLLOCK, NAE, University of Michigan, Ann Arbor (until June 2015)
PAUL R. SACKETT, University of Minnesota, Minneapolis
MARK S. SQUILLANTE, IBM Research, Yorktown Heights, New York
WILLIAM J. STRICKLAND, HumRRO, Alexandria, Virginia
STEVEN TADELIS, University of California, Berkeley

Staff

MICHELLE K. SCHWALBE, Senior Program Officer, *Study Director*
SCOTT WEIDMAN, Board Director
CHERIE CHAUVIN, Senior Program Officer
LINDA CASOLA, Senior Program Assistant
RODNEY N. HOWARD, Administrative Assistant
BETH DOLAN, Financial Manager
JONATHAN YANGER, Research Associate
CLAIRE JI, Christine Mirzayan Science and Technology Policy Graduate Fellow

[1] NAE, National Academy of Engineering.
[2] NAS, National Academy of Sciences.

COMMITTEE ON APPLIED AND THEORETICAL STATISTICS

CONSTANTINE GATSONIS, Brown University, Providence, Rhode Island, *Chair*
DEEPAK AGARWAL, LinkedIn, Mountain View, California
KATHERINE BENNETT ENSOR, Rice University, Houston, Texas
MONTSERRAT (MONTSE) FUENTES, North Carolina State University, Raleigh
ALFRED O. HERO III, University of Michigan, Ann Arbor
DAVID M. HIGDON, Social Decision Analytics Laboratory, Biocomplexity Institute of Virginia Tech, Arlington
ROBERT E. KASS, Carnegie Mellon University, Pittsburgh, Pennsylvania
JOHN LAFFERTY, University of Chicago, Illinois
XIHONG LIN, Harvard University, Cambridge, Massachusetts
JOSÉ M.F. MOURA, NAE, Carnegie Mellon University, Pittsburgh, Pennsylvania
SHARON-LISE T. NORMAND, Harvard University, Cambridge, Massachusetts
GIOVANNI PARMIGIANI, Harvard University, Cambridge, Massachusetts
ADRIAN RAFTERY, NAS, University of Washington, Seattle
LANCE WALLER, Emory University, Atlanta, Georgia
EUGENE WONG, NAE, University of California, Berkeley

Staff

MICHELLE K. SCHWALBE, Senior Program Officer
RODNEY N. HOWARD, Administrative Assistant
LINDA CASOLA, Senior Program Assistant
BETH DOLAN, Financial Manager

Acknowledgments

This report has been reviewed in draft form by individuals chosen for their diverse perspectives and technical expertise. The purpose of this independent review is to provide candid and critical comments that will assist the institution in making its published report as sound as possible and to ensure that the report meets institutional standards for objectivity, evidence, and responsiveness to the study charge. The review comments and draft manuscript remain confidential to protect the integrity of the deliberative process. We wish to thank the following individuals for their review of this report:

Brenda Dietrich, NAE,[1] IBM,
Alexis Fink, Intel,
Christine Fox, Johns Hopkins University Applied Physics Laboratory,
Ronald D. Fricker, Jr., Virginia Polytechnic and State University,
Henry J. Hatch, NAE, U.S. Army (retired),
Carl Hoffmann, Human Capital Management and Performance LLC,
Ronald C. Kessler, NAS[2]/NAM,[3] Harvard Medical School,
Philip Neches, NAE, Teradata,
Stephen M. Pollock, NAE, University of Michigan, and
Michael G. Rumsey, U.S. Army Research Institute (retired).

[1] National Academy of Engineering.
[2] National Academy of Sciences.
[3] National Academy of Medicine.

Although the reviewers listed above have provided many constructive comments and suggestions, they were not asked to endorse the views presented in the report, nor did they see the final draft of the report before its release. The review of this report was overseen by Robert F. Sproull, University of Massachusetts, Amherst, NAE, who was responsible for making certain that an independent examination of this report was carried out in accordance with institutional procedures and that all review comments were carefully considered. Responsibility for the final content of this report rests entirely with the committee and the institution.

The committee would like to thank the following individuals for providing input to this study:

Joseph Adams, Institute for Defense Analyses,
Amy Alrich, Institute for Defense Analyses,
Jeff Angers, Office of the Assistant Secretary of the Army, Manpower and Reserve Affairs,
Beth Asch, RAND Corporation,
Stephanie Barna, Office of the Under Secretary of Defense (Personnel & Readiness),
Mark Breckenridge, Defense Manpower Data Center,
Iveta Brigis, Google People Operations,
Kelly Buck, Defense Manpower Data Center,
Roll Buller, RAND Corporation,
Samuel Buttery, Naval Postgraduate School,
John Carbone, Office of the Assistant Secretary of the Army, Manpower and Reserve Affairs,
Linda C. Cavalluzzo, Center for Naval Analyses,
Ray Conley, RAND Corporation,
Elia Crisman, Office of the Chief of Naval Operations,
Matthew Davis, Defense Manpower Data Center,
Bruce Doll, Defense Health Agency,
Michael Dominguez, Institute for Defense Analyses,
Colin Doyle, Institute for Defense Analyses,
Mark Engelbaum, U.S. Air Force, Manpower, Personnel and Services,
Alexis Fink, Intel,
Colum Foley, IBM Smarter Workforce,
Christine Fox, Johns Hopkins Applied Physics Laboratory,
Peggy Golfin, Center for Naval Analyses,
Jay Goodwin, U.S. Army Research Institute,
Kristy Gore, RAND Corporation,
David Gregory, Center for Naval Analyses,
Jane Handley, RAND Corporation,
Kim Hepner, RAND Corporation,

Carl Hoffmann, Human Capital Management and Performance, LLC,
Stanley Horowitz, Institute for Defense Analyses,
Mike Housman, Cornerstone OnDemand,
David Hudak, Office of the Secretary of Defense,
Magdi Kamel, Naval Postgraduate School,
David Knapp, RAND Corporation,
Jennifer Kurkoski, Google People Operations,
Lance Landower, RAND Corporation,
Marco Laumanns, IBM Research,
Paul B. Lester, U.S. Army, Research Facilitation Laboratory,
Scott Lutterloh, Department of the Navy, Manpower and Reserve Affairs,
Arun Maiya, Institute for Defense Analyses,
Philip Major, Institute for Defense Analyses,
Bill Mason, Department of Defense, Office of Information,
Sean McDonald, Office of the Chief of Naval Operations,
Laura Miller, RAND Corporation,
Dave Nap, RAND Corporation,
David Nicholls, Institute for Defense Analyses,
Mary Kate Norton, Google People Operations,
Laura Odell, Institute for Defense Analyses,
Peter Olsen, U.S. Air Force, Manpower, Personnel and Services,
Julie Pechacek, Institute for Defense Analyses,
Jeff Petersen, Center for Naval Analyses,
Christopher Rice, Intel,
Susan Rose, Institute for Defense Analyses,
Mark Rosen, Center for Naval Analyses,
Johannes Royset, Naval Postgraduate School,
Jackie Ryan, IBM Smarter Workforce,
Richard Schaeffer, Office of the Assistant Secretary of Defense,
Ed Schmitz, Center for Naval Analyses,
Brandon Shapiro, Institute for Defense Analyses,
Yu-Chu Shen, Naval Postgraduate School,
Max Simkoff, Cornerstone OnDemand,
Don Simrod, Center for Naval Analyses,
Cara Sims, RAND Corporation,
Allison Taylor, Institute for Defense Analyses,
Fred Thompson, Office of the Under Secretary of Defense (Personnel & Readiness),
George Tolis, Institute for Defense Analyses,
Mike Walker, United States Military Academy at West Point,
Kayla Williams, RAND Corporation, and
John D. Winkler, RAND Corporation.

Contents

Summary

The Office of the Under Secretary of Defense (Personnel & Readiness), referred to throughout the report as P&R, is responsible for the total management of all Department of Defense (DoD) personnel, including recruitment, readiness, and retention. This mission requires extensive data, a large number and variety of complex analyses, and access to skilled workers to extract meaningful information to guide DoD personnel and readiness policies. With the advent of newer sources of data, such as social media and modern data analytics, P&R has the opportunity to exploit new tools that may produce more powerful analyses and improve the effectiveness and efficiency with which it accomplishes its mission. However, cultural and technological challenges exist and must be addressed, including the following: improving data access and sharing while ensuring proper privacy protection, enhancing analytic methods, and improving workforce education. An important step in addressing these challenges is developing a data and analytics framework, taking into account current and desired capabilities and addressing barriers accordingly. This National Academies of Sciences, Engineering, and Medicine report of the Committee on Strengthening Data Science Methods for Department of Defense Personnel and Readiness Missions offers suggestions on which data analytics capabilities could be targeted and which considerations to keep in mind to advance the framework for these capabilities. The study's full statement of task is shown in Box S.1.

This report considers data science in its broadest sense: a multidisciplinary field that concerns technologies, processes, and systems to extract knowledge and insight from data and to support reasoning and decision

BOX S.1
Study Statement of Task

An ad hoc committee will develop a roadmap and implementation plan for the integration of data analytics in support of decisions within the purview of the Department of Defense's Office of the Under Secretary of Defense (Personnel & Readiness).

making under various kinds of uncertainty. There are two primary aspects of interest within the field of data science, namely (1) the management and processing of data and (2) the analytical methods and theories for descriptive and predictive analysis and for prescriptive analysis and optimization. The first aspect involves data systems and data preparation, including databases and warehousing, data cleaning and engineering, and some facets of data monitoring, reporting, and visualization. The second aspect involves data analytics and includes data mining, text analytics, machine and statistical learning, probability theory, mathematical optimization, and visualization of results.

Currently, analyses developed to support P&R are often disjointed, one-time efforts that respond to immediate questions and may lack any plan for future use of their data or methods. A comprehensive data and analytics framework, properly implemented, could add coherence to this work, expanding the types of questions that P&R can quickly examine, reducing the cost of analyses, improving the reliability of findings, and better informing policy decisions. While developing this framework, both the short-term and long-term needs of the Secretary of Defense and the responsibilities of P&R should be considered.

The Force of the Future initiatives[1] being pursued by Secretary of Defense Ashton Carter aim to make the DoD workforce more equitable, efficient, and flexible through a number of efforts such as increasing the interchange of personnel with the civil sector, offering more family-friendly benefits, changing how military personnel are promoted, and improving the opportunities for civil service personnel. One aspect of this would be the establishment of an Office of People Analytics to better harness DoD's big data capabilities in the service of managing personnel talent. This would be done by increasing the understanding of personnel characteristics and analyzing how policy or environmental changes will affect the performance or composition of the workforce. The development of a data and analytics

[1] Please see Chapter 2 for a more detailed discussion of Force of the Future.

framework could revolutionize how data and analytics are used by P&R while contributing to the goals of the current Force of the Future initiative.

> **Finding:** Despite the substantial amount of data available on DoD personnel, the data may not be appropriate for DoD's analytic tasks, or they may necessitate considerable investment in constructing the variables of interest.

> **Finding:** Analyses developed to support the Secretary of Defense are often disjointed, one-off activities undertaken to respond to immediate questions and may lack a plan for future use of data or analytic methods.

> **Finding:** The reuse of operational data for analytic purposes can expose issues in data collection, recording, transmission, cleaning, coding, and loading. Problems are often not detected until the point of analysis, when anomalies crop up in results.

> **Recommendation 1:** The Office of the Under Secretary of Defense (Personnel & Readiness) should develop a data and analytics framework, and a strategy to implement that framework, that addresses both the principal outcomes of its responsibilities and the short-term and long-term needs of the Secretary, based on the findings, recommendations, and discussions outlined in this report and in the Force of the Future proposals.

Developing a data and analytics framework is a complex task, with many components that need to be addressed both individually and systematically. Data need to be easily accessible and shared across groups in a way that reduces the hurdles currently faced when researchers and analysts seek to find or share data while ensuring proper privacy and security protections. Analytic methods available to P&R need to be expanded to enable stronger and more rapid responses to significant P&R research and analysis questions. Prescriptive methods that would allow P&R to better assess alternatives and recommend actions could be used more extensively. The workforce that P&R relies on for its analytics also needs to be improved, which is a challenge facing organizations worldwide. Each of these components is briefly described in the following sections.

The following sections also discuss potential short-, medium-, and long-term goals to help move P&R in the direction of developing a data and analytics framework. Data quality and sharing can be improved immediately, while data science methods can be enhanced in the medium term and data science education strengthened in the long term.

IMPROVE DATA QUALITY AND SHARING

Collections of traditional administrative and transactional data used for P&R missions continue to grow owing to improved technical abilities to track and store data and an increased interest in capturing data that could provide meaningful insights. While the sheer quantity of data is growing in most domains, the importance of P&R's mission makes it essential that these data are better understood and utilized. Steps have been taken to simplify and unify available data—such as the development of the Defense Manpower Data Center (DMDC), a unified personnel file, and the Civilian Personnel Data System—and these efforts have greatly enhanced the ability of the Office of the Secretary of Defense (OSD) to understand the behavior of its personnel and the effects of its policies. Still, there are benefits to be gained by enabling deeper and richer collection and sharing of data, including improved force readiness, better allocation of funds, and a more agile and adept workforce.

One important step toward improving the usefulness of data would be to add new fields and formats to personnel files that would improve the productivity of P&R's analyses. This would require P&R to work with other organizations to identify the most useful such fields and formats. One particular technical need is to ensure that future records are capable of including unstructured data and free-form text, and that methods are available to search and combine such information reliably.

Another important need is to enable greater data sharing among the Services[2] and between P&R and the Services. This goal is challenging because data definitions can differ, software may not match up, and business practices may vary. These differences of practice reflect the separate histories of the Services. Moreover, there can be substantive reasons for the differences, rooted in variations in policies and practices across the Services. However, data sharing provides clear benefits, such as exposing incomplete or flawed data. Along these lines, there is also a need for standardization in the way data users can report problems with data collections and channel those problems back to the data providers when appropriate. Challenges of data sharing and repurposing are significant; in particular, different definitions and formatting of data complicate data merging and linking, making it difficult to bring to bear multiple databases and the additional insights they represent to inform studies. The reuse of administrative data for analytic purposes can also expose issues in data collection, recording, transmission, cleaning, coding, and loading. Problems with data are often not detected until the point of analysis, when anomalies crop up in results. In addition, there are significant up-front hurdles to identifying, accessing,

---

[2] Throughout the report, the term "Services" is used to refer to the U.S. military Services—namely, the Air Force, Army, Coast Guard, Marine Corps, Navy, National Guard, and Joint Chiefs.

and assembling the data needed to pursue desired data analyses. These hurdles can discourage decision makers from asking for data analyses and researchers from offering them.

DMDC has developed the Person-Event Data Environment (PDE), which is designed to bring data together in a unified and secure system where researchers can conduct analyses easily and in a privacy-preserving fashion. The PDE is a positive step in attempting to make data more easily accessible. However, technical and cultural challenges (such as possible data reidentification and other privacy compromises), a slow and complicated approval process to gain access, lengthy reviews for data import and export, limited computational capabilities, concerns about data quality and comprehensiveness, and concerns about data ownership rules pose a significant deterrent to utilizing the PDE. In addition, it is not clear that the architecture scales up in such a way that it can serve all of P&R's needs, and forcing analysts to work through the PDE personnel, who then must work through the data owners, may represent a barrier between the analyst and the raw data. The substantial efforts undertaken by PDE personnel to prepare the data for linkage are not transparent and may inadvertently impact the results of analysis.

> **Finding:** The existence of DMDC and a unified personnel file has greatly enhanced OSD's ability to understand the behavior of its personnel and to refine its policies so as to improve both retention and performance. The creation of the Civilian Personnel Data System was a similar achievement.

> **Finding:** There are benefits to be gained by enabling deeper and richer collection and sharing of data, which support a richer picture of the individual. This could in turn allow for better matching of personnel to the needs at hand (e.g., with regard to desired data skills, language proficiencies, and experiences), improved identification of at-risk servicemembers, enhanced management of the force in terms of retention and training, and many other benefits.

> **Finding:** The challenges of data sharing and repurposing are significant; in particular, different data definitions and formatting complicate data merging and linking. Business practices (e.g., methods, procedures, processes, and rules) vary from Service to Service and from one database to another.

> **Finding:** Enhanced data sharing within DoD, across agencies, and with the research community at large could promote the creation of new statistical methods, tools, and products.

Finding: The existence of alternative data sources, such as social media, especially when they are tied to extensive information about individuals, may deliver deep insights relevant to the mission of P&R. Owing to concerns about privacy and appropriateness and to the difficulty of ensuring statistical validity, further pursuit of this path requires careful consideration and additional research.

Recommendation 2: The Office of the Under Secretary of Defense (Personnel & Readiness) should investigate the feasibility of exploiting alternative data sources to augment traditional methods for measuring collective sentiment, evaluating recruitment practices, and classifying individuals (for creditworthiness, perhaps, or for battle-readiness). Hand in hand with this effort there should be an investigation into privacy technology appropriate for these scenarios for data use.

Recommendation 3: The Office of the Under Secretary of Defense (Personnel & Readiness) should identify incentives to enhance data sharing and collection, such as the following:

- Tracking usage of data by source in repositories such as the Person-Event Data Environment and periodically reporting back to data providers on usage (e.g., number of uses, who the users are, the nature of the study, or analysis the data contributed to);
- Providing incremental funding on contracts that involve data collection and organization to cover the costs of archiving and documenting the data for other users; and
- Giving preference to projects for constructing or redesigning operational data systems that include explicit functionality to support data sharing.

Recommendation 4: The Office of the Under Secretary of Defense (Personnel & Readiness) should leverage opportunities to improve access, including better reuse of prior data, tools, and results, and should investigate incentives to increase interagency and inter-Service data sharing.

Recommendation 5: The Office of the Under Secretary of Defense (Personnel & Readiness) should establish a working group with representation from the Services and other elements of the Department of Defense, as appropriate, to

- Identify productive new fields and formats for personnel files, such as enabling the inclusion of unstructured data and free-form text in future records;

- Identify opportunities for data sharing between Services and the Office of the Under Secretary of Defense (Personnel & Readiness) and within Services and lower barriers to such sharing;
- Work with organizations that provide operational data or collect them for analysis to improve data quality by providing standard ways for data users to report problems with data collections and channel those reports back to data providers when appropriate;
- Clarify self-reporting rules and practices;
- Identify legal and regulatory barriers to the appropriate and responsible sharing of data; and
- Examine new hardware and software architectures that facilitate data access and data management.

Finding: The development of the Person-Event Data Environment is a positive step in making some data more easily accessible. However, certain technical and cultural factors deter the use of this tool.

Benefits

- Spreads the overall cost of data acquisition, cleaning, ingestion, and linking;
- Reduces time for researchers identifying and downloading data, since they work on it in situ;
- Aims to improve handling of sensitive data;
- Monitors data usage; and
- Creates a group that supports users with data and tool issues.

Drawbacks

- Sensitive personally identifiable information is susceptible to reidentification and other privacy compromises such as revelation of sensitive traits or attributes.
- Linkage attacks—innocuous data in one data set used to identify a record in a different data set containing both innocuous and sensitive data—can be carried out via external data sets brought into the PDE by researchers.
- Review processes are lengthy for access to some data.
- Delays in the review process for export of analysis results pose a deterrent to publication and peer review.
- The hurdles to become a PDE user mean that the current user community is much smaller than intended.
- Some users have been limited by the computational power, memory, and tools of the current installation.
- The PDE does not solve completeness and quality issues in the underlying data sources.

- There does not exist a systematic mechanism for reporting data problems.
- Some PDE users say they have been given conflicting statements about the ownership of external data uploaded into the PDE.

Recommendation 6: The Defense Manpower Data Center should assess how well the Person-Event Data Environment is working and whether it is serving its intended community. In doing so, the center should consider taking the following steps to improve the usability of the Person-Event Data Environment and enhance its value:

- Assess if current privacy and security policies are adequate, taking into account modern methods of attack and sources of auxiliary information that can aid in these attacks, such as multiple releases of statistics and data sets (Ganta et al., 2008), linkage attacks that make use of public sources (Sweeney, 1997; Narayanan and Shmatikov, 2008), and chronological correlations with public sources (Calandrino et al., 2011).
- Analyze data usage information, both for privacy and determining value of assets.
- Do a better job of establishing and defining a user community for knowledge sharing. This includes improving relationships with the federally funded research and development centers doing work for the Department of Defense and determining which researchers would benefit from the capabilities of the Person-Event Data Environment.
- Remove unnecessary barriers for researchers to gain access to the system.
- Enhance computational power, memory, and tools.
- Respond to concerns about the quality and comprehensiveness of available data.
- Develop an explicit process for reporting data problems.
- Clarify data ownership rights to external data that are uploaded and merged.
- Assess protocols for accessing personally identifiable information.
- Review approval process for exporting analysis results.
- Consider widening access to the data and/or rebalancing Institutional Review Board requirements by establishing a differentially private interface.[3]

[3] For a discussion of differential privacy, see Chapter 5.

When conducting analyses relating to personnel data, it is essential that privacy, confidentiality, and fairness be considered as primary factors rather than, as is too often the case, left as an afterthought secondary to an analyst's findings. The current privacy and confidentiality protections in place with government databases rest heavily on Institutional Review Board (IRB) supervision. However, significant barriers arise in the overreliance on IRB reviews. Researchers often face multiple IRB reviews and re-reviews throughout a single study, which can significantly slow the research and analysis process and add months or years to the time it takes before researchers can access DoD data. This impedes their ability to respond to policy needs in a timely manner while also doing little to stop data reidentification and other compromises of sensitive personal information.

Finding: Reviews by multiple Institutional Review Boards can significantly slow down the research process and add months or years to the time it takes for researchers to have access to DoD data. This creates a serious problem for responding to policy needs in a timely manner.

Recommendation 7: In order to support timely and efficient research, the Office of the Under Secretary of Defense (Personnel & Readiness) should encourage streamlining of Institutional Review Board processes that involve multiple organizations—for example, federally funded research and development centers and the Department of Defense.

Recommendation 8: The Department of Defense should carry out research on the feasibility of differential privacy methods for its personnel analytics. These methods could reduce the need for Institutional Review Board oversight.

Recommendation 9: The Department of Defense should consider adopting or adapting the privacy and governance structure developed by the Office of Management and Budget for civilian statistical agencies. In particular, the department should follow the guidance on use of administrative records and establishing of statistical units under the Confidential Information Protection and Statistical Efficiency Act for both military and civil service personnel. In doing so, the department should examine the applicability of Fair Information Practice Principles in the treatment of Defense Manpower Data Center data.

Recommendation 10: The Defense Manpower Data Center, in its role as steward of the Person-Event Data Environment, should consider ways to adapt and use privacy and governance practices that the Office of Management and Budget has created for civilian use.

ENHANCE ANALYTIC METHODS

While comprehensive and reliable data are essential in informed decision making, they would be of little use without advanced analytic capabilities. New methods of analyzing data are increasingly available, and many cutting-edge approaches are ready to be more thoroughly applied for P&R missions. Data analytics are often categorized as descriptive analytics, predictive analytics, or prescriptive analytics. These categories are defined in Box S.2.

There are a number of opportunities to use other prescriptive analytic techniques beyond those currently being used for P&R missions. In industry, for example, studies addressing the problem of retention typically start with a statistical analysis (predictive analytics) of the entire workforce, looking to identify characteristics of personnel most likely to leave. This analysis estimates losses for the different groups of personnel and also produces models that estimate changes in losses for each group as a function of the amount of additional compensation provided to that group. Then, mathematical optimization under uncertainty methods (prescriptive analytics) are developed, incorporating these predictive models, to determine the compensation to be offered to reduce losses (increase retention) for each group in the workforce to best match demand. This decision-making optimization also takes into account the various trade-offs among other workforce policy levers such as hiring (recruiting) and reskilling (training) to best match demand.

Several hurdles need to be overcome to exploit these new opportunities. The previously discussed improvements to data access and sharing are a first step. The best analytic methods can be accessed by enhancing training in the workforce and by building a data analytics center such as the Office of People Analytics, proposed in the Force of the Future initiatives.

> **Finding:** A wide range of problems are being addressed for P&R using data analytic techniques and the rich data sources discussed in this report. These are often applied in response to specific questions but are not incorporated into a long-term plan.
>
> **Finding:** Turnkey personnel analytic solutions and currently commercially available software are unlikely to meet P&R's needs.
>
> **Recommendation 11:** The Office of the Under Secretary of Defense (Personnel & Readiness) should assess which predictive and prescriptive analyses would benefit its mission over the longer term, taking into account its understanding of which specific decisions could, if evaluated by applying more powerful data and/or methods, better enable

BOX S.2
Data Analytics Definitions

Data analytics definitions are from Lustig et al. (2010)[a] and Dietrich et al. (2014).[b]

Descriptive analytics. Examples include frequencies, distributions, tabulations, and visualizations.

- Set of technologies and processes that use data to understand and analyze an organization's performance measures.[a]
- Reporting what happened, analyzing contributing data to determine why it happened, and monitoring new data to determine what is happening now.[b]

Predictive analytics. Examples include classification, linear and nonlinear regression, data mining, text analysis, machine learning, Bayesian methods, and simulation.

- The extensive use of data and mathematical techniques to uncover explanatory and predictive models of an organization's performance representing the inherent relationship between data inputs and outputs/outcomes.[a]
- Techniques such as statistics and data mining to analyze current and historical information to make predictions about what will happen in the future, typically producing both a statement of possible events that could occur and the associated probabilities of their occurrence.[b]

Prescriptive analytics. Examples include stochastic models of uncertainty, mathematical optimization under uncertainty, and optimal solutions.

- A set of mathematical techniques that computationally determine a set of high-value alternative actions or decisions given a complex set of objectives, requirements, and constraints, with the goal of improving organizational performance.
- Analytics methods that recommend actions with the goal of finding an action (or set of actions) that will maximize the expected value (e.g., utility) associated with the outcome.

---

[a] I. Lustig, B. Dietrich, C. Johnson, and C. Dziekan, 2010, The analytics journey, *INFORMS Analytics Magazine*, November/December.

[b] B.L. Dietrich, E.C. Plachy, and M.F. Norton, 2014, *Analytics Across the Enterprise: How IBM Realizes Business Value from Big Data and Analytics*, Indianapolis: IBM Press.

the Department of Defense to prepare for future demands it may face. Some possible steps that might follow include these:

- Emphasizing the use of prescriptive analytics in conjunction with predictive "what if" scenarios;

- Enhancing prescriptive analytics usage and disseminating best practices across the entire department; and
- Adapting the prescriptive analytics methods successfully used in the private sector for workforce and talent management.

Controlled experiments offer an opportunity to test potential policy solutions and provide additional data when needed. They can be used for a variety of areas important to P&R and can be particularly helpful in buttressing conclusions that contradict accepted conclusions.

Finding: The Department of Defense does not routinely employ controlled experiments to understand causes and effects of the Office of the Under Secretary of Defense (Personnel & Readiness) policies—for example, revisions to enlistment standards or choices affecting family welfare—to judge whether they produce the intended effects and provide benefits that justify their costs.

Recommendation 12: To the extent feasible and relevant, the Department of Defense should conduct carefully structured experiments to test the efficacy of policy.

STRENGTHEN DATA SCIENCE EDUCATION

A skilled workforce that can apply state-of-the-art methodology and adapt to the quickly evolving data analytics domain is essential. OSD would benefit if P&R strengthened the data analytics expertise of a portion of its staff, both military and civilians. Such background would allow these specialized staff to answer immediate questions quickly, to be better-informed consumers of external analyses, and to better integrate analyses into policy decisions. Such expertise would also help to transfer best practices and skills across silos within the P&R enterprise.

Finding: Based on its collective experience with seeing data science mature in other organizations, the committee's judgment is that P&R's skills, depth, and resources in data analytics are not sufficient to recognize the full range of analytics opportunities and to implement these methods to better support decision making. It is always problematic to leverage scattered pockets of data science expertise, so raising the general level of awareness and skill would be more effective.

Recommendation 13: The Office of the Under Secretary of Defense (Personnel & Readiness) should create greater awareness of data science methods and disseminate them more thoroughly to its personnel to increase the general understanding of data science and the benefits of its use.

Recommendation 14: The Office of the Under Secretary of Defense (Personnel & Readiness) should enhance education in data science for its personnel, including civil service employees. This education could range from short courses in specific techniques for personnel who already have the requisite foundational knowledge, to overview seminars for managers who need to be acquainted with what their analytical staff can undertake, to formal degree programs, whether at Department of Defense or civilian universities.

REFERENCES

Calandrino, J.A., A. Kilzer, A. Narayanan, E.W. Felten, and V. Shmatikov. 2011. "You might also like": Privacy risks of collaborative filtering. Pp. 231-246 in *Proceedings of the 2011 IEEE Symposium on Security and Privacy.* May 22-25.

Ganta, S.R., S. Kasiviswanathan, and A. Smith. 2008. Composition attacks and auxiliary information in data privacy. *Proceedings of the 14th ACM SIGKDD International Conference on Knowledge Discovery and Data Mining.* http://www.cse.psu.edu/~ads22/privacy598/papers/gks08.pdf.

Narayanan, A., and V. Shmatikov. 2008. Robust de-anonymization of large sparse datasets. Pp. 111-125 in *Proceedings of the 2008 IEEE Symposium on Security and Privacy.* doi:10.1109/SP.2008.33.

Sweeney, L. 1997. Weaving technology and policy together to maintain confidentiality. *Journal of Law, Medicine and Ethics* 25(2-3):98-110.

1

Introduction

OVERVIEW

The Office of the Under Secretary of Defense (Personnel & Readiness), referred to throughout the report as P&R, is responsible for the total force[1] management of all Department of Defense (DoD) components, including the recruitment, readiness, and retention of personnel. Its work and policies are supported by a number of organizations both within DoD, including the Defense Manpower Data Center (DMDC), and externally, including the federally funded research and development centers (FFRDCs) that work for DoD: the Institute for Defense Analyses (IDA), Center for Naval Analyses (CNA), and RAND. P&R must be able to answer questions for the Secretary of Defense such as how to recruit people with an aptitude for and interest in various specialties and along particular career tracks and how to assess on an ongoing basis servicemembers' career satisfaction and their ability to meet new challenges. P&R must also address larger-scale questions, such as how the current realignment of forces to the Asia-Pacific area and other regions will affect recruitment, readiness, and retention.

New analytical methods for obtaining insight are of critical importance because of challenges such as maintaining a leaner but high-quality force, ensuring "reversibility" after the current personnel drawdown, managing escalating health-care costs, and identifying other cost issues early. Cur-

[1] An aggregation of military personnel, weapon systems, equipment, and necessary support, or combination thereof (DoD, *Dictionary of Military Terms*, 2015).

rently, personnel costs, including for training, health care, and compensation, absorb more than half of the budget for the DoD (CBO, 2012).[2]

While DoD makes use of large-scale data and mathematical analysis in intelligence, surveillance, reconnaissance, and elsewhere—exploiting techniques such as complex network analysis, machine learning, streaming social media analysis, and anomaly detection—these skills and capabilities have not been applied as well to the personnel and readiness enterprise. In the private sector, momentum has been building behind such efforts in recent years, and considerable work has been done for several decades. As noted in "They're Watching You at Work," from *Atlantic Monthly*, "The emerging practice of 'people analytics' is already transforming how employers hire, fire, and promote. . . . Predictive statistical analysis, harnessed to big data, appears poised to alter the way millions of people are hired and assessed" (Peck, 2013). Efforts have also been made in predictive (statistical) and prescriptive (optimization) analytics to address workforce (hiring and retention) management, skill and talent (readiness) management, and human capital resource allocation by the private sector (Hu et al., 2007; Cao et al., 2011; Hoffmann et al., 2012; Dietrich et al., 2014, Chapter 2). Even earlier related work can be found in, for example, White (1970), Bartholomew (1973), Vajda (1978), Gael (1988), and the references therein.

Secretary of Defense Ashton Carter has started several initiatives aimed at improving the Force of the Future, in part through the improved use of data analysis and updated technology (Garamone, 2015; Tilghman, 2015). The proposed opportunities for paid tuition, installation of new offices and occupations, and creation of new technologies all have the potential to improve recruitment, placement, and retention.

P&R has traditionally collected opinion data from surveys and focus groups, and those data underpin the analyses that have been performed. A large amount of administrative data are also available and hold great potential for further exploration. There has been a proliferation of available data and databases in the DoD enterprise functions, along with great advances in methods for gathering, storing, and accessing big data, and there have been advances in analytic techniques useful in working with large data sets. With these raw materials, data science can be applied to improve the efficiency and effectiveness of the DoD enterprise functions and to improve DoD's planning in general.

There are great opportunities for new types of data collection and analysis. Some of the challenges to be overcome include the following:

- Modern tools and concepts of data science are evolving rapidly. Creating more effective analyses based on the vastly enlarged data

[2] This amount does not include benefits provided by the Department of Veterans Affairs.

space available today and by newer methods of modeling, analysis, and optimization will require the use of the latest capabilities.
- Data often reside in different forms, and in various places, and are too often weak and perhaps not readily shared.
- Integration of new data and analytical capabilities with traditional data and results from earlier analyses can be problematic.
- New ideas may be needed to make optimal use of distributed data, distributed computing, and distributed analysis, because it may no longer be practical or desirable for all of these assets to be colocated.
- New workforce skills are needed within DoD to exploit these opportunities.

DoD is not the only entity facing this challenge. However, its mission is essential to the welfare of the United States, and there are considerable opportunities to advance its analytic capabilities and unique opportunities for data collection and long-term tracking of large numbers of personnel.

STUDY ORIGIN AND APPROACH

This National Academies of Sciences, Engineering, and Medicine report, sponsored by DoD and the National Security Agency, addresses the statement of task for the Committee on Strengthening Data Science Methods for Department of Defense Personnel and Readiness Missions, presented in Box 1.1.

The study committee held 6 meetings of the full committee to collect information and deliberate. The open session presentations at these meetings are listed in Appendix C. The committee also conducted 14 site visits with the following organizations and groups: Office of the Under Secretary of Defense (Personnel & Readiness); Office of the Chief of Naval Operations, Manpower and Personnel; Office of the Assistant Secretary of the Army, Manpower and Reserve Affairs; Air Force Office of Manpower, Personnel and Services; RAND; CNA; IDA; Naval Postgraduate School; Defense

BOX 1.1
Study Statement of Task

An ad hoc committee will develop a roadmap and implementation plan for the integration of data analytics in support of decisions within the purview of the Department of Defense's Office of the Under Secretary of Defense (Personnel & Readiness).

Manpower Data Center; Defense Readiness Reporting System; Google's People Analytics group; Intel's Talent Intelligence and Analytics group; Cornerstone OnDemand; and Workday. These visits allowed the committee to discuss data science challenges and lessons learned in more detail.

In the course of its meetings and site visits, the committee raised questions such as the following:

- What is the state of the art of data analytics for enterprise functions outside DoD? How might the newest methods be applied to recruiting, readiness, and retention? Which methods are appropriate for the special circumstances within DoD (such as not being able to recruit into service at advanced rank from outside)? What DoD responsibilities or problems are not covered by commercial technologies? How might the commercial technologies need to be adapted to the DoD environment?
- In which areas of importance to P&R (and the enterprise side of DoD more generally) are there likely to be benefits from new efforts in data analytics? What is the nature and size of those benefits? What might DoD be able to do that cannot be accomplished within the current analytical environment?
- How do the current DoD information technology and analytics infrastructure (both hardware and software) and practices compare with the state of the art? What are the benefits and costs of transitioning from the current infrastructure and data architectures? What types of human capital would be required to make the transition and then to operate in the new environment(s)?
- Which sources of external information (e.g., social media or demographic data) could complement the internal sources of information that P&R already collects and maintains? Which of these have potential for shortening problem detection and correction? Which of these could improve current estimation and projection products? What types of research and analysis are required to develop analytical capabilities that better leverage existing sources of information and enable appropriate use of new sources? How would current analytical approaches and skill sets need to change?
- How can P&R best leverage data analytics capabilities that already exist in other parts of DoD (e.g., in components dealing with R&D and with intelligence operations)?
- How can tracking, evaluation, and performance estimation capabilities be improved within an organization?
- What are the difficulties in modeling and analyzing unprecedented situations, such as shifts in deployment policy, including the location and duration of assignments?

- What would be a suitable overall strategy for transitioning P&R to support and exploit new data analytic capabilities? In general terms, what estimated level of effort and resources would be required?

While part of P&R's mission relates to health care–related policies, the committee did not explicitly examine health care data and their unique considerations. Such a discussion would have strong links to (1) the challenges of developing and exploiting electronic health care records, which is a topic that extends well beyond DoD and cannot be usefully examined through any one study, and (2) the challenges of strengthening coordination between DoD and the Veterans Administration, which likewise cannot be usefully examined through any one study. Instead, the committee focused on how to use administrative, transactional, and unconventional data more effectively.

ORGANIZATION OF THE REPORT

While the primary audience for this report is individuals affiliated with P&R, the committee made an effort to have it be useful for any interested party looking for information on developing and employing data science methods. For those unfamiliar with P&R, Chapter 2 provides a brief overview, including its organizational structure, objectives, functions, and key responsibilities. Building on the information presented in Chapter 2, Chapter 3 discusses data and analysis capabilities available to P&R, both internal to DoD and external, utilizing resources such as the FFRDCs. That chapter also discusses how these data and analyses inform P&R decision making. For a more technical discussion of the mathematics and analyses utilized in current data science methods, Chapter 4 outlines the importance and challenge of data preparation and important descriptive, predictive, and prescriptive data science methods that may benefit from further exploration. A reader who is interested in the legal and privacy issues associated with implementation of data science methods can be directed to Chapter 5 for a discussion of privacy and confidentiality concerns relating to conducting analyses with personnel data. Chapter 6 discusses the commercial state of the art in human resources analytics, both in terms of available commercial products and practices used across industry. Chapter 7 provides the committee's findings and recommendations, including opportunities and possible approaches for P&R to improve data quality and sharing through improved planning and coordination between data holders, streamlined Institutional Review Boards for outside researchers, and improved access with relevant privacy considerations; to enhance the use of data science methods by developing a longer-term approach to integrating advanced predictive and prescriptive analytics and utilizing controlled experiments to test policy changes; and to enhance data science education within the P&R analytic workforce.

REFERENCES

Bartholomew, D.J. 1973. *Stochastic Models for Social Processes*. 2nd edition. London: Wiley.

Cao, H., J. Hu, C. Jiang, T. Kumar, T.-H. Li, Y. Liu, Y. Lu, S. Mahatma, A. Mojsilovic, M. Sharma, M.S. Squillante, and Y. Yu. 2011. OnTheMark: Integrated stochastic resource planning of human capital supply chains. *Interfaces* 41(5):414-435.

CBO (Congressional Budget Office). 2012. *Costs of Military Pay and Benefits in the Defense Budget*. November.

Dietrich, B.L., E.C. Plachy, and M.F. Norton. 2014. *Analytics Across the Enterprise: How IBM Realizes Business Value from Big Data and Analytics*. Indianapolis: IBM Press.

Gael, S. 1988. *Job Analysis Handbook for Business, Industry and Government*. New York: Wiley.

Garamone, J. 2015. Carter details Force of the Future initiatives. *DoD News*. November 18. http://www.defense.gov/News-Article-View/Article/630400/carter-details-force-of-the-future-initiatives.

Hoffmann, C., E. Lesser, and T. Ringo. 2012. *Calculating Success: How the New Workplace Analytics Will Revitalize Your Organization*. Boston: Harvard Business Review Press.

Hu, J., B.K. Ray, and M. Singh. 2007. Statistical methods for automated generation of service engagement staffing plans. *IBM Journal of Research and Development* 51(3):281-293.

Peck, D. 2013. They're watching you at work. *Atlantic Monthly*. November 20. http://www.theatlantic.com/magazine/archive/2013/12/theyre-watching-you-at-work/354681/.

Tilghman, A. 2015. 'Force of the Future': Career flexibility, fewer moves. *Military Times*. August 30. http://www.militarytimes.com/story/military/careers/2015/08/28/force-future-report-ash-carter-review/32476549/.

Vajda, S. 1978. *Mathematics of Manpower Planning*. New York: Wiley.

White, H.C. 1970. *Chains of Opportunity: System Models of Mobility in Organizations*. Boston: Harvard University Press.

2

Overview of the Office of the Under Secretary of Defense (Personnel & Readiness)

INTRODUCTION

The Office of the Under Secretary of Defense (Personnel & Readiness), referred to throughout the report as P&R, is responsible for myriad functions that cover over 3 million personnel (including servicemembers, civil service employees,[1] and contractors), retirees, veterans, and dependents. At its core, P&R is responsible for establishing policies for recruitment, placement, and retention of 1.3 million active duty military servicemembers, 1.1 million National Guard and Reserve members (DoD, 2015a), and 750,000 appropriated fund civil service employees (Defense Civilian Personnel Advisory Services, DoD Demographics as of September 30, 2014) who work for the Department of Defense (DoD).[2] P&R makes a large number of consequential decisions and must assess how best to approach those decisions. In the process of daily operations, it generates and manages large quantities of data that are used, although not to full potential in the committee's view.

This chapter outlines the responsibilities of P&R within the three general categories of readiness and force management, manpower and reserve

---

[1] This amount does not include benefits provided by the Department of Veterans Affairs. Civil service employees are defined as "all appointive positions in the executive, judicial, and legislative branches of the Government of the United States, except positions in the uniformed services" (5 U.S.C. § 2101). While civil service employees work in most government agencies, this report refers only to those employed by DoD.

[2] In addition, there were 125,000 nonappropriated fund civil service employees and 53,000 local national employees.

affairs, and health affairs, along with some of the decisions that are made and the research that is produced in these categories.

P&R STRUCTURE AND ORGANIZATION

As laid out in the U.S. Code,[3] the Under Secretary of Defense (Personnel & Readiness) is responsible for military readiness; total force management; military and civil service personnel requirements; military and civil service personnel training; military and civil service family matters; exchange, commissary, and nonappropriated fund[4] activities; personnel requirements for weapons support; National Guard and reserve components; and health affairs (10 U.S. Code §136). P&R advises the Secretary of Defense and creates policies, procedures, and standards for recruitment; assignment to positions; promotion within the ranks; retention of personnel; and the transition to veteran status. In addition, it is responsible for administering benefits such as health care.

P&R is thus responsible for policy and oversight in the following areas:

- Staffing decisions for both servicemembers and civil service employees;
- Recruiting standards for both servicemembers and civil service employees;
- Selection criteria for military recruits;
- Selection criteria for civil service positions;
- Job assignment;
- Compensation standards for servicemembers;
- Training and education programs for personnel;
- Promotion criteria for servicemembers and civil service employees;
- Security clearance determinants;
- Access to DoD buildings and locations;
- Process for and time of transitions out of the armed forces;
- Provisions for mental and physical health care to servicemembers, civil service employees, retirees, and dependents;
- Suicide prevention among servicemembers;
- Provisions for the needs of families;
- Responses to problematic behavior;

[3] Title 10, Section 136 (10 U.S. Code §136).

[4] Nonappropriated funds are defined as funds derived from sources other than congressional appropriations and commissary surcharge funds, primarily from the sale of goods and services to servicemembers, DoD civil service personnel, and family members of both who are used to support or provide morale, welfare, and recreation programs (DoD 4105.67).

- Provisions for health and retirement benefits to separated servicemembers;
- Responses to congressional requests for information; and
- Prediction of future needs in all of these areas.

P&R is responsible for a broad set of outcomes regarding DoD personnel and their dependents. For many of those outcomes, its instrument is the policy pronouncement or regulatory mechanism, with the military Services (Army, Navy, Air Force, and Marine Corps) responsible for actually carrying out the policy.[5] P&R, however, can judge the success of the policy (or lack thereof) only through empirical monitoring of the outcomes, which provides the basis for adjustments. (In the extreme, P&R can seek amended or new statutory authority.) Such monitoring also provides the basis for reporting on outcomes to the Secretary of Defense, the President, and the Congress.

For some issues, P&R, not the military departments, is also the administrative agent. Examples include the Defense Commissary Agency, the TRICARE health care program, the Federal Voting Assistance Program, the Defense Travel Office, and the Defense Manpower Data Center (DMDC). As these examples illustrate, P&R holds responsibility for execution as the supervisor of a defense agency (or activity). And in emergent situations the Under Secretary's immediate staff may have some responsibility for policy execution (as was the case, for example, in creating a Family Support Center in the aftermath of the 9/11 attack on the Pentagon, or in dealing with certain aspects of sexual assault). Even more so in these circumstances, the Under Secretary's team needs an empirical monitoring capability, both to assess the effectiveness of policy and to ensure it is implementing those policies well.

Oversight of support programs for servicemembers and veterans, as well as their families, is also central to the mission of P&R. P&R encourages partnerships among other agencies, governments, and communities (Office of the Under Secretary of Defense [Personnel & Readiness], 2015).

The objectives of P&R shift as guidance from Congress and the Secretary of Defense change over time. Each year, for example, the Congress

[5] The relationship between the military Services and the Office of the Secretary of Defense (and, therefore, P&R) is complex. At one level, P&R establishes policies, and the Services carry out or comply with those policies. At the same time, the Services establish their own policies (which may not be inconsistent with P&R's policies) within their mission requirements. For example, P&R may require the Services to collect and report for P&R use certain data on each servicemember. The Services will do that, but they may also collect—and may or may not report to P&R—other data on those same servicemembers that a Service determines are necessary for its operations. While cognizant of these Service data collections, the committee focused its task on data available to P&R.

authorizes a maximum size for the active force and appropriates funds to pay that size force. The Services (Army, Navy, Air Force, and Marine Corps), with P&R oversight, then establish recruiting targets to reach—but not exceed—that force size, based on expected losses during the year. Similarly, P&R may establish targets for highly qualified enlistees, based on models that consider recruiting costs, training costs, and expected proficiency of the force.

The Under Secretary of Defense (Personnel & Readiness) supervises three Assistant Secretaries with the following responsibilities: (1) readiness and force management, (2) manpower and reserve affairs, and (3) health affairs. The Under Secretary also oversees the Defense Human Resources Activity (DHRA) and the Executive Director Force Resiliency. Each of these areas is discussed further in the following subsections, and a current organization chart is shown in Figure 2.1.

READINESS AND FORCE MANAGEMENT

The Assistant Secretary of Defense for Readiness and Force Management assesses whether the current servicemembers and civil service employees are adequately trained and the units to which they are assigned are prepared for their responsibilities (Deputy Chief Management Officer memorandum of October 29, 2015). This assistant secretaryship is a new position, and it remains to be seen how it will focus its activities.

Manpower and Reserve Affairs

The two major tasks this office undertakes are developing personnel policies for civil service employees and servicemembers. The Office of Civilian Personnel Policy establishes policies related to human resource issues and develops strategies and procedures for the effective management of DoD's civilian workforce, and may operate in partnership with the Office of Personnel Management (OPM). Within the Office of Civilian Personnel Policy, the Civilian Personnel Management Service provides assistance to the military departments and defense agencies on matters of employment and leadership. DoD's foreign national employment program also relies upon the Office of Civilian Personnel Policy for guidance related to employment of such persons. Additionally, the Office of Civilian Personnel Policy is responsible for oversight of the nonappropriated fund personnel system (Office of the Under Secretary of Defense [Personnel & Readiness], 2015).

The Office of Military Personnel Policy is responsible for establishing policy affecting all Services and for overseeing Service-specific policies to ensure that they are aligned with both DoD policy and congressional direction. Since 1965, the Office of Military Personnel Policy has overseen

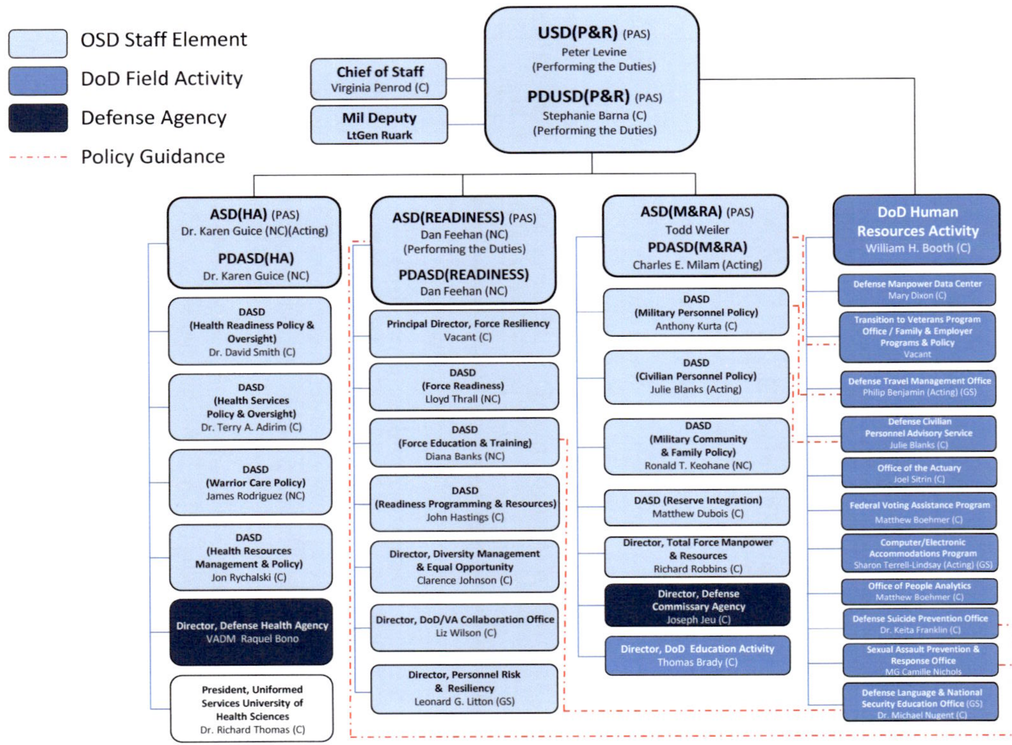

FIGURE 2.1 Organization and leadership chart for the Office of the Under Secretary of Defense (Personnel & Readiness), as of March 2016. NOTE: Acronyms are defined in Appendix A. SOURCE: Office of the Under Secretary of Defense (Personnel & Readiness).

production of a quadrennial report on military compensation (see, for instance, DoD, 2012a). These reports are a large undertaking[6] that lays out issues regarding cash and noncash compensation for the active and reserve components of the military. Within the Office of Military Personnel Policy, the Office of Accession Policy establishes policy, planning, and review in order to maintain the force levels established by Congress. This office oversees the recruitment and processing of "accessions," the military's term for individuals who join the military. In addition to overseeing enlisted servicemembers, it oversees officer accession policy through, for instance, the service academies and the Reserve Officer Training Corps.

The Assistant Secretary for Manpower and Reserve Affairs, through the above-mentioned Office of Accession Policy, oversees the recruitment of enlisted servicemembers, a task that involves administering more than 1 million cognitive and noncognitive ability tests a year and more than 250,000 physical exams (USMEPCOM, 2015), with the goal of enlisting approximately 200,000 active duty individuals per year across the Services. The armed forces have traditionally judged recruits based on their cognitive, physical, and educational characteristics, as well as their behavioral records (Sackett and Mavor, 2003). To this end, the Military Entrance Processing Command administers the Armed Services Vocational Aptitude Battery (ASVAB) and medical exams, and conducts background checks to allow the Services to determine which recruits to admit and in what occupation each recruit should be trained (DoD, 2012b). Recruits take these tests at one of 65 military entrance processing stations (MEPSs) and more than 300 mobile sites (USMEPCOM, 2015). In 2014, these sites administered nearly 400,000 entrance tests.

DoD administered an additional 682,000 ASVABs in high schools during the 2014-2015 school year, with the goal of providing career advice to students while at the same time providing information to recruiters (USMEPCOM, 2015). The ASVAB includes multiple-choice questions in nine areas: general science, arithmetic reasoning, mathematics knowledge, word knowledge, paragraph comprehension, electronics information, automotive and shop information, mechanical comprehension, and assembling objects.[7] Data gathered in this process are documented in the annual reports *Population Representation in the Military Services* (DoD, 2016, for example).

Some Services have recently begun considering applicants' results on the Tailored Adaptive Personality Assessment System (TAPAS), developed to measure characteristics that reflect resilience and persistence, to screen

---

[6] These reports typically require 2 years to complete at an estimated cost of $9 million per report.

[7] The ASVAB website, www.official-asvab.com/index.htm, was accessed January 5, 2016.

candidates not just on the basis of traditional cognitive aptitude screening criteria, but also on their personality traits (e.g., openness, conscientiousness, extraversion, agreeableness, and emotional stability) (Heffner et al., 2011; Stark et al., 2014). DoD has plans to assess this new screening tool with a longitudinal study of its use (Sheftick, 2014), and there is hope that the ASVAB and TAPAS assessments and other measurements of performance potential can be further refined to aid in selecting members for particular units as opposed to general recruitment only (NRC, 2013).

There are nearly as many reserve and National Guard members (1.1 million) as there are active duty members (1.3 million) in the armed forces (DoD, 2015a). The Office of Manpower and Reserve Affairs has responsibility for six branches of reserve components in DoD, one each attached to the Army, Navy, Air Force, and Marine Corps, as well as National Guard units associated with the Army and the Air Force (Reserve Affairs, 2015). The office also facilitates the incorporation of the reserves into the armed forces through the following programs: readiness, training, and mobilization; materiel and facilities; and family and employer programs and policies (Reserve Affairs, 2015).

A significant challenge in managing the reserves is figuring out how best to change their status from reserve to active duty—that is, how to mobilize them. Since 2001, P&R has overseen the mobilization of more than 800,000 reserve force members (Reserve Affairs, 2015).

HEALTH AFFAIRS

The Under Secretary also oversees the Assistant Secretary of Defense for Health Affairs under Directive Number 5136.01, who, in turn, oversees the Military Health Service (MHS), which cares for more than 9.6 million beneficiaries with a budget of more than $50 billion (DoD, 2014a; Military Health Service, 2012). The health service determines how best to provide for the health care needs of servicemembers, their families, civil service employees, and military retirees. As do other health care providers, it balances providing health care while minimizing cost (DoD, 2014a).

Health Affairs is also responsible for policy relating to the health status of current servicemembers, the health care of retirees, and the health care of families of both servicemembers and retirees. It not only oversees the TRICARE system, which operates somewhat like a health insurance provider, but it is also responsible for policy at health care facilities operated by the Services (e.g., hospitals and clinics), caring directly for servicemembers, retirees, and their families. Extensive data on health care utilization are collected through TRICARE and these military treatment facilities.

To manage the system, Health Affairs needs to know the types of health care that are needed, the types of providers that are available, and how those providers are distributed geographically. It also must work to foster wellness among the individuals who are served by the system. In addition, the Assistant Secretary of Defense for Health Affairs has mandated a focus on the health consequences of deployment, particularly mental health. Servicemembers are required to complete the Postdeployment Health Assessment (PDHA) immediately after returning from deployment and the Postdeployment Health Reassessment (PDHRA) 3-6 months later (Appenzeller et al., 2007; AFHSC, 2009).

DEFENSE HUMAN RESOURCES ACTIVITY

The Under Secretary also oversees the Defense Human Resources Activity (DHRA), which is responsible for overseeing and carrying out a variety of personnel-related activities in DoD, including the following:

- Overseeing the integration, support, and training of women in the military through the Defense Department Advisory Committee on Women in the Services. Within this arena, it must evaluate whether current policies and programs are effective and efficient.
- Working with employers of reserve members and overseeing the transition into and out of active status for reserves through the Employer Support of the Guard and Reserve.
- Providing information on regional and language needs in the Defense Language and National Security Education Office.
- Managing the DMDC, which is the central repository of DoD human resource information.
- Advising the Under Secretary on issues related to compensation and benefits in the Office of the Actuary.
- Providing leadership in civil service employee management through the Defense Civilian Personnel Advisory Service.
- Overseeing the Federal Voting Assistance Program, which manages the Uniformed and Overseas Citizens Absentee Voting Act.
- Coordinating DoD commercial travel through the Defense Travel Management Office.
- Determining how DoD responds to suicide, in terms of both preventing and responding to these acts, in the Defense Suicide Prevention Office.
- Preparing servicemembers and their families for separation that occurs during deployment through the Family and Employer Programs and Policies Office.
- Offering assistance via policy and program oversight to servicemembers entering the civilian workforce through the Transition to Veterans Program Office.

- Managing the Computer/Electronic Accommodations Program, which funds an accommodation program for employees with disabilities.
- Connecting National Guard and Reserve members, families, and communities with resources during deployment through the Yellow Ribbon Reintegration Program (YRRP) (DoD, 2015b).

In 2005, DHRA established the Sexual Assault Prevention and Response Office (SAPRO), which oversees Service responses when servicemembers are perpetrators or victims of sexual assault and develops programs aimed at preventing assault (DoD, 2015c). DoD reports on sexual assaults within the military, using administrative data on sexual assault complaints and how those complaints are resolved (DoD, 2015d) as well as periodic surveys.

Executive Director Force Resiliency

The recent reorganization of P&R created this new office to coordinate efforts on suicide, sexual assault, collaboration with the Department of Veterans Affairs (VA), and diversity. SAPRO and the Defense Suicide Prevention Office will continue as part of DHRA, but the Executive Director will be responsible for creating a cohesive set of policies and actions across the range of these challenging problems. The Executive Director for Force Resiliency oversees the Office of Diversity Management and Equal Opportunity (ODMEO), the DoD/VA Collaboration Office, and Personnel Risk Reduction.

ODMEO is responsible for the development and execution of diversity management and equal opportunity policies and programs affecting all DoD personnel. The Director of ODMEO also provides supervision, direction, and policy guidance of the Defense Equal Opportunity Management Institute (DEOMI) at Patrick Air Force Base, Florida, which conducts equal opportunity and equal employment opportunity training, education, and research.

The DoD/VA Collaboration Office is responsible for managing the transition of approximately 200,000 active duty servicemembers to veterans annually (Joint Chiefs of Staff, 2014) and the transition as Reserve Component members are released from active duty. In doing so, these servicemembers may move from the purview of the Department of Defense to that of the VA.

The Office of Personnel Risk Reduction manages the accident reduction and safety portfolios and the Drug Demand Reduction Program. The office is focused on the operational safety arena, which incorporates military training, aviation operations, deployment, operational employment (combat), human factors (high-risk behaviors), wellness of civil service employees and servicemembers (to include deterring drug abuse), and leadership engagement to promote awareness and culture change.

Main Deliverables

P&R produces dozens of recurring research reports based on administrative data. Some of the most notable are *Population Representation in the Military Services* (DoD, 2016), classified quarterly readiness reports to Congress,[8] the *Defense Manpower Requirements Report* (DoD, 2014b), and the *Quadrennial Quality of Life Review Report* (DoD, 2009). P&R also relies on congressional testimony as a means of disseminating information. In addition, P&R produces annual reports on desired and actual recruitment and retention.

P&R DECISIONS

The principal outcomes for which P&R is responsible can be grouped under six key areas.[9] Each of these is discussed in more detail in Chapter 3, and some of the ways in which advanced data analytics could contribute to each are discussed in Chapter 7:

1. Ensuring DoD can recruit, train, motivate, and retain the necessary numbers of qualified personnel.
2. Creating incentives that guide the department to an optimal mix of personnel.
3. Ensuring DoD creates a force that is ready to carry out directed actions.
4. Influencing DoD's decisions that affect the shape of military careers.
5. Ensuring the services provided to support DoD personnel and their dependents are properly structured.
6. Anticipating and responding to sensitive behavioral issues.

FORCE OF THE FUTURE

Early into his term as Secretary of Defense, Ashton Carter committed himself to rethinking DoD policies for both military and civil service employees (Carter, 2015a). This reexamination was motivated by a concern

[8] For a discussion of the contents of these reports, see GAO (2013).

[9] Grouping is necessary given the broad range of P&R's responsibilities. One sign of the breadth of that range is that DoD directive 5124.02 is the most frequently referenced by other directives; second most referenced is 5134.01 for the Under Secretary of Defense (Personnel & Readiness). Even if one conditions the search on "acquisition," it is the second most frequently cited.

that DoD is not attracting, managing properly, or retaining the talent it needs at this time. Following that speech, the then Acting Under Secretary of Defense (Personnel & Readiness), Brad Carson, spoke out frequently, raising fundamental challenges about civil service rules (e.g., authority to appoint) and military promotions (e.g., the up-or-out precept embodied in the Defense Officer Personnel Management Act).

An ambitious internal review process was launched, culminating in a set of far-reaching proposals in late summer 2015. For the military, these included replacing up-or-out with perform-or-out, whereby one might stay at the same military grade for extended periods of time, permitting greater lateral entry (the current paradigm starts virtually everyone at the bottom, except for professionals such as medical doctors, nurses, and lawyers) and adopting new benefits (such as a vast expansion of parental leave) to encourage more men and women to continue their service. For civil service employees, it was proposed that new hiring gateways be granted to DoD, and that non-bargaining unit employees be governed under the Secretary of Defense's authority (Title 10 of the U.S. code) via OPM (Title 5). New mechanisms for interchange with the civil sector were advanced and new oversight mechanisms were recommended (e.g., an Office of People Analytics).

A lengthy internal debate on these ideas ensued, focusing on their likely effects and costs (presumably also involving the OMB and White House staff). On November 18, 2015, Secretary Carter delivered a speech at George Washington University, "Building the First Link to the Force of the Future," focusing principally on permeability, announcing a series of internships and exchange-type programs, and pledging to continue to work on the larger issues (Carter, 2015b; DoD, 2015e). He later elaborated on some of these initiatives (Carter, 2016a; Carter, 2016b). An accompanying Fact Sheet was released describing a new Office of People Analytics (OPA) (DoD, 2015f), which will use big data analytics to better understand key components of servicemembers' career paths such as hiring, planning, and training. In its first year of operation, OPA plans to do the following:

- Appoint an acting director and acquire a team of highly-skilled data scientists and organizational psychologists with expertise in civilian workforce issues, military manpower issues, data storage, analytical and statistical methods, and social science research methods;
- Make an initial investment to purchase and assemble the supporting technology architecture, which requires modern and secure data storage, in-memory analytics, data analysis tools, and data visualization tools; and
- Convene an advisory board within a working group of stakeholders to categorize existing data sets, prioritize analytic projects, and

identify needs for new data sets (Mark Breckenridge and Kristin Williams, personal communication, April 19, 2016).

Once fully funded, OPA will explore ways that DoD can use analytics to better understand how policy or environmental changes affect the performance and composition of its workforce.

Some additional Force of the Future proposals concerning data and data management include these:

- *Examining ways to improve recruiting.* To avoid attrition costs to DoD, P&R would initiate and supervise a study that would reward military recruiters based on their recruits' performance during the initial enlistment term and basic training. This study, conducted by a federally funded research and development center, would be used to advise DoD on the factors driving poor recruitment outcomes. The study is scheduled to start before November 2016.
- *Implementing a web-based talent management system.* The Services would develop a database populated with input from individual servicemembers via a web-based system in an effort to better match members with their assigned positions. DoD believes such a system would better satisfy the needs of both DoD and the individual members.
- *Implementing exit surveys.* Exit surveys would become standard practice to better understand the reasons why individuals leave military service.
- *Updating and modernizing the retirement system.* DoD would continue its work on the Blended Retirement System to allow greater career path choices for current and future servicemembers (DoD, 2015d).

REFERENCES

AFHSC (Armed Forces Health Surveillance Center). 2009. Update: Deployment health assessments, U.S. Armed Forces, December 2008. *Medical Surveillance Monthly Report* 16(1):12-16.

Appenzeller, G.N., C.H. Warner, and T. Grieger. 2007. Postdeployment health reassessment: A sustainable method for brigade combat teams. *Military Medicine* 172(10):1017-1023.

Carter, A. 2015a. Remarks on the Force of the Future as Delivered by Secretary of Defense Ash Carter, Abington Senior High School, Abington, Pennsylvania. March 30. http://www.defense.gov/News/Speeches/Speech-View/Article/606658.

Carter, A. 2015b. Remarks on "Building the First Link to the Force of the Future" as Delivered by Secretary of Defense Ash Carter. George Washington University Elliott School of International Affairs, Washington, D.C., November 18. http://www.defense.gov/News/Speeches/Speech-View/Article/630415/remarks-on-building-the-first-link-to-the-force-of-the-future-george-washington.

Carter, A. 2016a. "Department of Defense Press Briefing by Secretary Carter on Force of the Future Reforms in the Pentagon Press Briefing Room." News transcript. January 28. http://www.defense.gov/News/Transcripts/Transcript-View/Article/645952/department-of-defense-press-briefing-by-secretary-carter-on-force-of-the-future.

Carter, A. 2016b. "Remarks on 'The Next Two Links to the Force of the Future.'" Speech transcript. June 9. http://www.defense.gov/News/Speeches/Speech-View/Article/795341/remarks-on-the-next-two-links-to-the-force-of-the-future.

DoD (Department of Defense). 2009. *Report of the 2nd Quadrennial Quality of Life Review.* Washington, D.C.: Department of Defense.

DoD. 2012a. *Report of the 11th Quadrennial Review of Military Compensation.* Washington, D.C.: Department of Defense.

DoD. 2012b. "United States Military Entrance Processing Command (USMEPCOM): DoD Directive 1145.02E." Washington, D.C.: Department of Defense.

DoD. 2014a. *Final Report to the Secretary of Defense: Military Health System Review.* Washington, D.C.: Department of Defense.

DoD. 2014b. *Defense Manpower Requirements Report.* Washington, D.C.: Department of Defense. http://prhome.defense.gov/Portals/52/Documents/RFM/TFPRQ/docs/F15%20DMRR.pdf.

DoD. 2015a. *2014 Demographics: Profile of the Military Community.* Washington, D.C.: Department of Defense. http://download.militaryonesource.mil/12038/MOS/Reports/2014-Demographics-Report.pdf. Accessed December 11, 2015.

DoD. 2015b. "Defense Human Resources Activity." http://prhome.defense.gov/DHRA. Accessed July 20, 2015.

DoD. 2015c. "Sexual Assault Prevention and Response: Mission and History." http://www.sapr.mil/index.php/about/mission-and-history. Accessed January 10, 2016.

DoD. 2015d. *Department of Defense Annual Report on Sexual Assault in the Military.* Washington, D.C.: Department of Defense.

DoD. 2015e. "Force of the Future: Maintaining Our Competitive Edge in Human Capital." Memorandum from Secretary of Defense Ash Carter. November 18. http://www.defense.gov/Portals/1/features/2015/0315_force-of-the-future/documents/SD_Signed_FotF_Memo_-_11.18.pdf.

DoD. 2015f. "Fact Sheet: Building the First Link to the Force of the Future." http://www.defense.gov/Portals/1/features/2015/0315_force-of-the-future/documents/FotF_Fact_Sheet_-_FINAL_11.18.pdf. Accessed January 4, 2015.

DoD. 2016. *Population Representation in the Military Services, Fiscal Year 2014.* Washington, D.C.: Office of the Assistant Secretary of Defense for Personnel and Readiness. https://www.cna.org/pop-rep/2014/. Accessed February 15, 2016.

GAO (General Accounting Office). 2013. *Military Readiness: Opportunities Exist to Improve Completeness and Usefulness of Quarterly Reports to Congress.* Washington, D.C.: General Accounting Office.

Heffner, T.S., R.C. Campbell, and F. Drasgow. 2011. *Select for Success: A Toolset for Enhancing Soldier Accessioning.* ARI Special Report 70. Ft. Belvoir, Va.: U.S. Army Research Institute for the Behavioral and Social Sciences.

Joint Chiefs of Staff. 2014. "Veteran Stereotypes: A Closer Look." White paper. October. http://www.jcs.mil/Portals/36/Documents/CORe/141024_veteran_stereotypes.pdf.

Military Health Service. 2012. *MHS: Health Care to Health: MHS Stakeholders Report.* Washington, D.C.: Military Health Service.

NRC (National Research Council). 2013. *New Directions in Assessing Performance of Individuals and Groups: Workshop Summary.* Washington, D.C.: The National Academies Press.

Office of the Under Secretary of Defense (Personnel & Readiness). 2015. "Office of the Under Secretary (Personnel & Readiness): About." http://prhome.defense.gov/About.aspx. Accessed July 21, 2015.

Reserve Affairs. 2015. "Reserve Affairs: The Office of the Assistant Secretary of Defense." http://ra.defense.gov/. Accessed July 21, 2015.

Sackett, P.R., and A.S. Mavor. 2003. *Attitudes, Aptitudes, and Aspirations of American Youth: Implications for Military Recruiting*. Washington, D.C.: The National Academies Press.

Sheftick, G. 2014. "Army Considering Major Changes for Recruiting." *Army News Service*. October 23. http://www.army.mil/article/136850/Army_considering_major_changes_for_recruiting/.

Stark, S., O.S. Chernyshenko, F. Drasgow, C.D. Nye, W.L. Farmer, L.A. White, and T. Heffner. 2014. From ABLE to TAPAS: A new generation of personality tests to support military selection and classification decisions. *Military Psychology* 26:153-164.

USMEPCOM (U.S. Military Entrance Processing Command). 2015. http://www.mepcom.army.mil/. Accessed November 24, 2015.

3

Personnel and Readiness Data and Their Use

INTRODUCTION

The Department of Defense (DoD) currently employs over 3 million personnel (DoD, 2015a; Defense Civilian Personnel Advisory Services, "DoD Demographics as of Sept 30, 2014") and operates and maintains over 800 military bases around the world (Vine, 2015). In order to efficiently manage and effectively use DoD resources to achieve its mission requirements, it is critical to obtain and understand data concerning the availability and readiness of both servicemembers and civil service employees, the condition of available and appropriate equipment, and the operating status of DoD installations. The Under Secretary of Defense (Personnel & Readiness) supports DoD's mission by advising it or developing policies and plans, as well as by monitoring the readiness of deployable combat and noncombat units. The Defense Manpower Data Center (DMDC) and its predecessor organization, the Manpower and Research Data Analysis Center, were created with the primary mission of supporting the information management needs of the Office of the Under Secretary of Defense (Personnel & Readiness), referred to throughout the report as P&R. Among other things, the DMDC collects and collates data from the military Services to inform P&R and DoD in their decision making. Before the creation of the DMDC in the early 1970s, there existed few cross-Service databases available to support policy or program decisions. Typically, when elements within the Office of the Secretary of Defense required cross-Service analysis, contractors requested data from the Services and assembled custom databases for those analyses. With the creation of DMDC, however, common data formats across the Services were able to be established in advance, and the data could be routinely provided for storage and maintenance—allowing analyses to be performed in a much more timely manner.

This chapter describes the data that DoD and P&R collect and the analytical capabilities available to these organizations, including the role of relevant federally funded research and development centers (FFRDCs). A brief description of the structure of P&R data follows. How these data and capabilities inform P&R decision making is then discussed. The findings and recommendations regarding DoD data capabilities can be found in Chapter 7 of this report.

CURRENT DATA AND ANALYSIS CAPABILITIES

Data Available to P&R

Under the supervision of the Defense Human Resources Activity (DHRA), DMDC collects personnel, training, financial, and other data for DoD. DMDC manages information relating to the retirement, health care, and other needs of personnel and their families. The center also collects and maintains much of the data collected by the Services and coordinates an overall database of 35 million individual records on servicemembers, employees, contractors, retirees, and family members.[1] DMDC operates in five areas—(1) decision support; (2) entitlements, benefits, and readiness reporting; (3) personnel identification, validation, and authentication; (4) enterprise integration; and (5) survey management—with data largely provided by the Services (DMDC, 2014). There are similar systems for civil service employee data, maintained as the Defense Civilian Personnel Data System,[2] and for the Reserve Component, maintained as the Reserve Components Common Personnel Data System (DoD, 2011). These record systems provide basic demographic data and a history of the individual's assignments; for military personnel, they include educational information and cognitive, physical, and moral aptitude test results.

DMDC is responsible for validating and issuing the Common Access Card (CAC), the identification card for servicemembers, civil service employees, and some contractors, which allows access to buildings and other spaces, as well as computer networks and systems (DMDC, 2016a). On average, 2.8 million CACs are issued annually, with approximately 3.4 million CACs in circulation (DHRA, 2009). DMDC also administers the following databases:

---

[1] For more information about DMDC, see https://www.dmdc.osd.mil/appj/dwp/dmdc_overview.jsp.

[2] For more information about this system, see http://cpol.army.mil/library/permiss/115.html.

- *Defense Enrollment Eligibility Reporting System (DEERS).* As a user-input database of personnel and benefits information on servicemembers, their families, and DoD civil service personnel and contractors, DEERS is used to determine who is eligible to receive benefits, decide who is to be issued CACs, help detect fraud and abuse in benefits programs, and answer other personnel and readiness questions. Enrollment in DEERS is a prerequisite for eligibility in many medical and dental benefits (such as TRICARE) (DMDC, 2016b).
- *Joint Personnel Adjudication System (JPAS).* JPAS tracks security clearances for all DoD personnel. It is divided into two systems: the Joint Adjudication Management System (JAMS) and Joint Clearance and Access Verification System (JCAVS). JAMS is responsible for recording changes in eligibility for security clearances, while JCAVS allows employers to view current security clearance statuses (Defense Security Service, 2016).
- *Automated Continuous Evaluation System (ACES).* ACES provides automated assessment of eligibility of DoD personnel to access classified information. This information includes responses to official questionnaires, personnel administrative data (e.g., demographic, employment, financial, educational, citizenship, and contact information), and criminal justice records. The automation of the database has streamlined the security clearance process, saving DoD both cost and time (PERSEREC, 2016).
- *Defense Biometric Identification System (DBIDS).* DBIDS is the current DoD identification system that combines biometrics and barcodes to identify and grant DoD personnel access to buildings and other secure locations, as well as to distribute departmental equipment and vehicles. Fingerprint or hand geometry is read and then combined with proper barcode readings before complete verification (Department of the Navy, Chief Information Officer, 2005).
- *Defense Incident-Based Reporting System (DIBRS).* Designed to meet DoD requirements for statutory reporting, DIBRS monitors, tracks, and organizes law enforcement information on crimes of interest and criminal investigations (DTIC, 2014).
- *Emergency Evacuation Tracking and Repatriation System.* Tracks persons who were evacuated while abroad owing to emergency situations and helps to recover the costs of such relocations (DPCLD, 2009).

Important additional sources of data are surveys undertaken by P&R, whether by DMDC acting for P&R, by a firm with which it contracts, or by one of its outside research organizations in support of the tasks it has been asked to undertake. For more than 15 years, DMDC has undertaken both one-off and ongoing surveys, the latter of which ask standardized questions at close intervals (e.g., once a year or even more frequently for military personnel). Recent examples include the Survey of Active Duty Spouses,[3] Gender Relations Surveys (for active duty,[4] reserve component,[5] and Service academies[6]), Survivor Experience Survey,[7] Workplace and Equal Opportunity Survey of Reserve Component Members,[8] and the Quickcompass of Financial Issues.[9] Besides eliciting demographic data from the respondents, these surveys seek to gather reports on specific phenomena (such as sexual assault) and on the outlook of the target personnel community toward their responsibilities and continued willingness to undertake them (such as questions about stress or intention to reenlist).

P&R also has available a variety of databases that derive from its management responsibilities. Prominent among these are the data sets it maintains on health care, whether provided in military facilities (overseen by the Defense Health Agency) or purchased from the private sector via the TRICARE contract program, which now accounts for well over half of all medical care that DoD finances. Likewise, in administering the Military Entrance Processing Stations (MEPSs)—which, as the name indicates, test incoming enlisted personnel—P&R accumulates test data for all potential military recruits.

Additional databases were established over time to collect essential data to address specific challenges. One such challenge is the possibility of operating more economically by using civil service employees in positions now filled with military personnel, who have higher lifetime costs for DoD. Executive agencies are required to conduct an annual inventory of the commercial and the inherently governmental activities being performed by federal employees.

---

[3] See http://download.militaryonesource.mil/12038/MOS/Surveys/ADSS1201-Briefing-Support-Deployment-Reintegration-PCS-WellBeing-Education-Employment.pdf.

[4] The website for the Gender Relations Surveys for active duty servicemembers is https://www.dmdc.osd.mil/appj/dwp/rest/download?fileName=WGRA1201_TabsVolume.pdf&groupName=pubGenderActive, accessed January 5, 2016.

[5] The website for the Gender Relations Survey for the reserve component is https://www.dmdc.osd.mil/appj/dwp/rest/download?fileName=WGRR1201_TabsVolume.pdf&groupName=pubGenderReserve, accessed January 5, 2016.

[6] See http://www.sapr.mil/public/docs/reports/MSA/APY_14-15/SAGR_DoD2015_FocusGroupReport.pdf.

[7] See http://www.sapr.mil/public/docs/reports/FY14_POTUS/FY14_DoD_Report_to_POTUS_Annex_2_DMDC.pdf.

[8] The website for the Workplace and Equal Opportunity Survey of Reserve Component Members is https://www.dmdc.osd.mil/appj/dwp/rest/download?fileName=WEOR1101_TabVolume.pdf&groupName=pubOpSurResComp, accessed January 5, 2016.

[9] The website for the Quickcompass of Financial Issues Surveys is https://www.dmdc.osd.mil/appj/dwp/rest/download?fileName=QCFIA1301_TabVolume.pdf&groupName=pubFinSurAD, accessed January 5, 2016.

This report, referred to as the "Inherently Governmental and Commercial Activities Inventory," looks at every position in DoD from the standpoint of whether it needs to be filled by a servicemember or a civil service employee.

Likewise, concerns in both Congress and the Executive Branch over the accuracy of reporting on the readiness of military units to carry out their assignments led in the last decade to establishing the Defense Readiness Reporting System (DRRS). A federation of existing databases, DRRS requires commanders—starting at the four-star level of the Combatant Commands—to periodically rate the ability of their units to carry out every mission-essential task in the operational plans, with the scoring responsibility cascading down the chain of command to the brigade level in the Army, for example, and analogously in the other Services (U.S. Army, 2011). To illuminate the reasons for readiness conditions, the system allows the user to query the supporting databases in the federated system (e.g., the database on equipment condition).

Beyond the large databases of this sort, P&R either assembles or has access to a variety of more limited databases, typically created to deal with a specific responsibility. Sometimes these are in the form of reports, whether recurring or one-time; however, a time series can be constructed only to the extent that the necessary archive of past submissions has been maintained. One of the most important one-time sources of data is the submission required annually for DoD's future resource planning process, the Program Objective Memorandum, usually known by its acronym, POM.

The Army Analytics Group and DMDC implemented a Person-Event Data Environment (PDE) in 2006 in an attempt to centralize portions of DoD health, military service, and demographic data into an electronic repository, with improved data security and accessibility (Vie et al., 2013). This capability was designed to make it easier to analyze topics such as unit readiness, recruitment trends, retention rates, and other key issues involving DoD personnel. The PDE is a first attempt to connect these disparate data.

While the data held within the PDE are neither classified nor secret, the PDE does hold sensitive information, such as medical records and fitness-for-duty reports (Vie et al., 2013). Therefore, maintaining the confidentiality of the data is critical. The PDE aims to do this by requiring researchers to apply for access to the data, explain what analyses are being conducted with the data, and access the PDE through a secure remote connection.

The committee met with PDE users from the FFRDCs, who noted that while the PDE vision holds promise for improving data access, some practical deficiencies exist, including a slow and complicated approval process for researchers trying to access the system; limited computational power, memory, and tools; concerns about the quality and comprehensiveness of data availability, difficulty uploading and merging external data, and ownership of data that is uploaded; and the extensive reviews required to access some personally identifiable data and to export analysis results.

Data Structure

Most of the data sources described were constructed originally for administrative purposes, not for policy analysis or research. The data elements in the sets are usually defined by the needs of the administrative process (e.g., billing records for purchased health care, which do not necessarily document the episode of illness or the patient's health status), and the data structure typically reflects the administrative process being supported. For example, to create a data file that tracks retention of military personnel, DMDC's records (which are typically monthly snapshots of Services personnel data) can be linked longitudinally, revealing if an individual serving at month t is still serving at month $t + 1$. Similarly, to evaluate the effectiveness of the Services' policies in granting waivers to entry standards for applicants who have disqualifying incidents in their background checks, P&R contracted for a study that combined recruit entry data with subsequent disciplinary actions, promotions, and other performance indicators (Putka et al., 2003; Putka, 2004). As these examples illustrate, the policy analyst must typically use more than one file and be willing to use proxy indicators for the underlying variables of interest in order to carry the inquiry forward.

The principal exceptions to this generalization are some of the data provided in response to a specific data call (e.g., for multiyear budget development) and the data collected by surveys. These efforts usually respond to an analytic issue that motivated the request or the collection of data. Nonetheless, those issues may be quasi-administrative in nature and driven by an immediate administrative need. For example, how does the beneficiary population view a specific benefit? Or what are the earnings from employment of military retirees? The data call or survey may not be designed to gather the variables that might explain *why* the benefit generates the observed response or the variables that explain the choices military retirees make in seeking and accepting civil employment.

Data Users

Federal staff and staff at the relevant FFRDCs are the principal users for all sources of data that P&R manages. DMDC's personnel records, for example, were used by P&R to answer Secretary Rumsfeld's questions about who was bearing the burden of military deployments to Iraq and Afghanistan during the extended operations in those countries. His concern reflected his interest in the all-volunteer force and a conviction that those burdens should be shared equitably. DMDC's data revealed a pattern of utilization that failed to meet his standard and led to extended discussion with military leaders about corrective actions (Baiocchi, 2013). Based on these discussions, Army and Marine Corps leadership took action that resulted in an improvement to the balance of personnel used in deployments (CRS, 2012).

There are 10 independent FFRDCs that work for DoD,[10] three of which conduct analysis and data collection principally related to personnel and readiness and utilize DMDC data: the National Defense Research Institute (NDRI), which is operated by the RAND Corporation; the Institute for Defense Analyses (IDA); and the Center for Naval Analyses (CNA).

NDRI analyzes issues related to human resources, including those relating to force management, readiness, support, and health care. It recently conducted, for example, a study on sexual assault and harassment that was based on a survey of 560,000 servicemembers (DoD, 2015b). It also recently reviewed personnel systems to examine how DoD matches its personnel to positions.

IDA examines defense policy and force planning in its Strategy, Forces, and Resources Division. In particular, it assesses issues related to organizational effectiveness and human capital management, as well as force structure and military capability alternatives. It recently completed a study, for example, of how to encourage personnel to learn to speak foreign languages and how to make sure that servicemembers who already have language and cultural skills are assigned the right locations. In this study, IDA argued that the military should not try to develop or to identify servicemembers with such skills but should instead contract out for these skills. It also conducted a study to project the psychological health care needs of active duty servicemembers.

CNA conducts research that evaluates workforce management and military readiness. Its researchers have, for example, developed simulation tools to project what positions are needed, how those positions will be staffed, and how the personnel in the positions are to be trained and educated. CNA also drafts congressionally mandated reports on servicemember characteristics (DoD, 2016).

Advisory panels to the government also draw on DMDC data. The Military Compensation and Retirement Modernization Commission used these data sources for crafting an alternative approach to military retirement and for testing its ability to replicate the current profile of military

---

[10] The 10 DoD FFRDCs are divided into R&D laboratories (Lincoln Laboratory, Software Engineering Institute, and Institute for Defense Analyses Communication and Computing Center), systems engineering and integration centers (The Aerospace Corporation and MITRE National Security Engineering Center), and study and analysis centers (CNA, IDA, RAND Arroyo Center, RAND NDRI, and RAND Project Air Force).

retention, and thus the experience levels the military would enjoy. Those estimates were produced by the FFRDCs that work for DoD, in this case, RAND's NDRI. Likewise, the IDA used the data sources of the Defense Health Agency to estimate for the Military Compensation and Retirement Commission the likely financial impact of a revised approach to the military health benefit.

DoD can and does turn to other analysis organizations besides the FFRDCs, including large for-profit consulting firms. In doing so, however, it must be especially vigilant about conflict of interest issues and analytic independence, which the FFRDCs are explicitly structured to protect. These firms may likewise employ the same data sources described here. For example, the Human Resources Research Organization (HumRRO) developed a cost-performance trade-off model for P&R to use in estimating the effects of recruiting budgets on the quality and job performance of new recruits (McCloy et al., 1992).

Groups with an interest in DoD policies and their implications may also seek to use these data sources. For example, the Services often use DMDC data in their Service-specific personnel research and analysis efforts. Independent scholars may also apply for access to these data sources.

As might be imagined, release of these data beyond the originating office is governed by the applicable statutes and federal policies, including those intended to protect human subjects. The federal staff enjoys the greatest access, as do advisory panels. The FFRDCs, by the terms of their charters and the restrictions placed upon them (such as specific policies for avoiding organizational conflicts of interest), may also be granted a high level of access, although not normally to personally identifiable information (records are often anonymized before being released to the FFRDCs). Although the FFRDC data requests respond to taskings from DoD, the granting of access to data may involve considerable negotiation and delay. Beyond the FFRDCs, sharing of these data with other organizations is more limited, although the Army and DMDC's PDE represents an effort to respond to requests for data in a manner that balances those requests against both privacy and government concerns.

FFRDCs house their own data in some cases, but usually such data collections are limited in scope and relate to ongoing or long-term studies. The majority of data utilized by FFRDCs are housed within DMDC and are often used in conjunction with external data such as those from the VA and the Bureau of the Census. Researchers are attempting to use the DMDC's PDE to access and utilize data but are encountering challenges, discussed earlier in this chapter and in Chapter 7.

P&R DECISION MAKING

Mission-Critical Decisions in Need of Improved Data Analytics

As mentioned in Chapter 2, P&R decision making is focused in six areas, each of which is discussed in this section:

1. Ensuring DoD can recruit, train, motivate, and retain the necessary numbers of qualified personnel;
2. Creating incentives that guide DoD to an optimal mix of personnel;
3. Ensuring DoD creates a force that is ready to carry out directed actions;
4. Influencing DoD's decisions that affect the shape of military careers;
5. Ensuring the Services supporting DoD's personnel are properly structured and provided; and
6. Anticipating and responding to sensitive behavioral issues.

Ensuring DoD Can Recruit, Train, Motivate, and Retain the Necessary Numbers of Qualified Personnel

P&R must ensure that DoD can recruit, train, motivate, and retain the necessary qualified personnel. For DoD civil service employees, P&R works in partnership with the Office of Personnel Management in accordance with Title V of the U.S. Code. However, there is a particular subset of DoD-employed civil service employees who are managed entirely within the department—for example, highly qualified experts and nonappropriated fund employees.[11]

P&R is responsible for ensuring the success of the all-volunteer force. Among other things it makes recommendations on pay policy and fringe benefits (which must be codified in statute) and on the setting of standards (physical, mental, and moral) for military service, and it participates in decisions on the host of factors that make up what one might call the "social compact" between DoD and those who choose to serve it.

For analysis of personnel issues—the recruit, train, motivate, retain function—P&R can turn to the Active Duty Military Personnel Master File maintained by DMDC, the Reserve Components Common Personnel Data System, and the Defense Civilian Personnel Data System. These record systems provide basic demographic data, educational information, and a history of individual pay grades and assignments. For civil service employees, the record includes training received, performance ratings and awards, and

[11] Nonappropriated fund employees are civilian employees who are paid from nonappropriated funds (10 U.S. Code § 1587).

disciplinary actions. The military personnel records include entrance exam results (including physical and cognitive aptitude test results) and military occupational specialty (defined for civil service employees by the position held and by occupational series). Because these records are continuously maintained, they can be used to analyze cohort behavior (e.g., retention of individuals from one year to the next or retirement from the civil service) and to relate those behaviors to characteristics such as gender, race, education, employment history, and income.

For a broader behavioral picture, P&R can turn to surveys undertaken by DMDC, principally of military personnel, both active and reserve components. For over 15 years, these surveys have been regularly fielded with standardized questions focused on retention and satisfaction with military life, plus usually at least one other topic of immediate interest. At longer intervals, surveys tailored to a major issue are fielded (e.g., employment of military spouses, earnings of military retirees, or sexual assault). Such focused surveys may also be undertaken by outside research organizations, including the FFRDCs, at the behest of P&R or one of the Services, which will also conduct their own surveys.

Contemporary budget pressures may be responsible for the marked reduction in the number and frequency of DMDC surveys beginning in 2010, relative to 2003-2009. This decrease makes the survey less useful as a predictive or leading-indicator tool. The Status of Forces survey of active duty personnel, for example, was typically taken two or three times a year in the earlier period. DMDC has also dropped its pursuit of QuickCompass surveys, which were intended to provide more rapid results than a standard survey instrument. Data analytics may turn out to be an economical substitute for the earlier survey approach.

Offsetting these reductions, DMDC launched its first longitudinal survey, the 2010 Military Family Life Project, focused on military family life and intended to measure the well-being of military spouses and families over time. This study consisted of four surveys given over a 2-year period, with every active duty spouse surveyed at least once to create a matched couple file. The Military Family Life Project also measured how families cope with relocations and deployments.

P&R and the military Services also use data from publicly available sources (e.g., the Employment Cost Index or the unemployment data published by the Department of Labor) to gauge the attractiveness of military service relative to civilian opportunities. P&R may collect more detailed data itself (e.g., its annual survey of prevailing wages, used to set blue-collar wages for both DoD and the federal government as a whole) or engage others to do so (e.g., tracking the propensity of American youths to volunteer for military service, ongoing since 1975[12]). Many of the data sources used to formulate decisions are also used to track and ensure that the decisions are robust.

P&R occasionally partners with other agencies to ensure its interests are reflected in the data they collect. Perhaps the most notable example is the National Longitudinal Survey of Youth 1979 (NLSY79), in which DoD financed an expanded military sample of the Department of Labor initiative, using it to create national norms for the Armed Services Vocational Aptitude Battery (ASVAB), the cognitive entrance examination for military service, and other purposes.[13]

Creating Incentives That Guide DoD to an Optimal Mix of Personnel

P&R helps create the incentives that guide DoD to an optimal mix of personnel, for which the headline issue is the mix of servicemembers, civil service employees, and contractors who staff the enterprise (see DoD Directive 1100.4). Because settling on the best mix of personnel types can engender considerable controversy, it is all the more important that the data on which these decisions are based be viewed as accurate and definitive. P&R's influence on overall personnel mix is often indirect, through budget decisions or planning rules that channel choices appropriately.

In contrast to the robust set of data sources available to P&R for the analysis of the "recruit, train, motivate, and retain" function, fewer data are available to analyze the "mix of personnel" issue. An important resource is the Fully Automated System for Classification (U.S. Army, 2012). P&R and allied offices (especially in the military Services) supplement this data source by commissioning analyses of specific personnel trade issues, often by the FFRDCs. Because of the interest in what the private sector should (or should not) do, these analyses have created considerable literature on public–private competitions (long an interest of the Office of Management and Budget under its Circular A-76).[14] That literature concludes that the

---

[12] Although the methodology changed between 1999 and 2001. See http://www.dtic.mil/dtic/tr/fulltext/u2/a416458.pdf.

[13] For a brief summary of NLSY79, see http://www.users.nber.org/~kling/surveys/NLSY79.html.

[14] OMB Circular A-76, Performance of Commercial Activities (05/29/2003), including technical corrections (OMB Memorandum M-07-02 (10/31/2006) and OMB Memorandum M-03-20 (08/15/2003)): https://www.whitehouse.gov/sites/default/files/omb/assets/omb/circulars/a076/a76_incl_tech_correction.pdf.

Examples of the "outsourcing" or "competitive sourcing" literature include Frances P. Clark et al., "The Impact of Large Multi-Function/Multi-Site Competitions," CNA, CRM D0008566.A2-Final, August 2003; Edward T. Morehouse, Jr., "Overview of Competitive Sourcing and Privatization, Framing the Issues for Army Environmental Cleanup," IDA, D-2359, February 2000; and Beth J. Asch and John D. Winkler, "Ensuring Language Capability in the Intelligence Community: What Factors Affect the Best Mix of Military, Civilians, and Contractors?" RAND, TR-1284-ODNI, 2013.

competitions typically generated savings for the federal government, often because a revised but streamlined government operation won (OMB, 2003). Government organizations often commission studies to estimate what staffing is required to undertake the assigned tasks.[15]

The private sector has invested heavily in relevant decision-making research areas. The problem of determining staffing requirements to undertake projects or tasks has been studied using predictive (statistical) analytics; for example, Hu et al. (2007) and Cao et al. (2011) apply statistical methods to historical data on business service engagements in order to infer the levels of staffing skills that have led to successful engagements. The optimal composition of the workforce over time has been considered through a combination of predictive analytics and prescriptive (optimization) analytics; this combination includes optimal decisions with respect to recruiting, retention, and training or retraining over time to realize the optimal workforce composition and address skill and talent shortages and surpluses; refer to Cao et al. (2011) and the corresponding examples in Appendix D. Lastly, the problem of sourcing—that is, choosing among different suitable resources to satisfy demand for a particular project or task—has been studied as a specific instance of a general class of dynamic resource allocation problems and addressed through approaches based on prescriptive analytics (refer to Gao et al., 2013, and the corresponding example in Appendix D). Hoffmann et al. (2012) and Dietrich et al. (2014) describe the use of data analytics to address various aspects of workforce and talent management in organizations.

Research in this area began in the 1970s. White (1970) studied workforce and talent management using models of mobility in organizations, including the notion of vacancy chains. Then, Bartholomew (1973) developed mathematical models of social phenomena and applied stochastic processes to workforce planning. Vajda (1978) also considered mathematical aspects of workforce planning. Gael (1988) described methods of analyzing jobs to meet specific situations and objectives, including job evaluation, wage incentives, job design, affirmative action, employee performance measurement, data collection techniques, and job diagnosis.

Ensuring DoD Creates a Workforce That Is Ready to Carry Out Directed Actions

P&R is responsible for helping to ensure that DoD creates a workforce that is ready to carry out the actions the President directs. While that

[15] See, for example, Gerald E. Cox, "Improving Cost Estimates in the Force Mix Allocation Process for the Active and Reserve Components," CNA, DRM-2012-U-003418-Final, February 2013; Thomas H. Barth et al., "Staffing Cyber Operations," IDA, NS D-5472, May 2015; and Terrence K. Kelly et al., "Stabilization and Reconstruction Staffing: Developing U.S. Civilian Personnel Capabilities," RAND, MG-580-RC, 2008.

responsibility is generally measured in terms of agreed indicators of unit performance (e.g., as instantiated in the DRRS), it also includes responsibility when units or individuals do not perform as expected. Its responsibility for readiness outcomes meant that P&R would play a significant role in the decision regarding which units would carry out operations in Afghanistan and Iraq (the so-called deployment orders process). Readiness also embraces the physical health of the workforce, with issues ranging from vaccination compliance to the stockpiling of medicines against pandemic disease.

DoD traditionally measured readiness through inputs to the Status of Resources and Training System (SORTS). It relied first on empirical measures (e.g., personnel fill) and then on an overall readiness judgment by commanders. The latter, of course, is subject to the same weakness as any judgmental score (such as manipulation in response to bureaucratic pressures),[16] and the former fails to address whether the unit can carry out its assigned missions. In the 1990s, Congress insisted on a review of readiness assessment bias, which helped justify the implementation of the revised system DRRS (Tillson et al., 2000). While DRRS does include SORTS data, it focuses on the mission-essential tasks associated with each operational plan and asks commanders—starting with the most senior and cascading down the chain—to evaluate the ability of their units to carry out those tasks. This process provides finer resolution and more nuanced data than were previously available. At the same time, DRRS provides access to other DoD data files that facilitate judging the reasonableness of those evaluations, the causes of any shortcomings, and the potential for remedial action.

Influencing DoD's Decisions That Affect the Shape of Military Careers

P&R plays a role in DoD's decisions that affect the shape of military careers—for example, promotion criteria and policies. The basic structure is prescribed by Congress through a combination of statutory authority and direction (e.g., the Defense Officer Personnel Management Act) and limits placed on certain choices (e.g., grade ceilings). Under Declarations of National Emergency, which have been in effect continuously since September 2001, Congress has given DoD broad authority to waive many of the statutory limitations, with the result that P&R must, at a minimum,

[16] In 1999, the *Washington Post* published an article entitled "Two Army Divisions Unfit for War," which told of both the 10th Mountain Division and the 1st Infantry Division receiving the lowest possible rating for unit readiness. In response to inquiries from Congress and the White House, the Army authorities said the primary reason for this rating was the peacekeeping mission in the Balkans, which required up to half of their troops. The Army went on to state that the two divisions in question were "more ready to fight than the new evaluation suggests."

advise on how that waiver authority should be used and often actually administer it.

Some Secretaries of Defense take an especially active interest in shaping military careers and enhancing the role that P&R is expected to play. During Secretary Rumsfeld's tenure, DoD proposed and persuaded Congress to make a number of statutory changes, including loosening restrictions on the length of military careers. Secretary Carter is likewise taking a strong interest in personnel issues, tasking his Acting Under Secretary of Defense (Personnel & Readiness) to consider sweeping changes in order to better manage DoD's talent and to attract new talent to its ranks. Although the shaping of military careers is importantly governed by statute, P&R is the steward of the process by which statutory changes are proposed. Change may arise from concerns of the military departments or Congress, or from the agenda of the Secretary of Defense or the President. Again, it is the personnel master data files, supplemented by survey responses and special studies, that give P&R insight into potential statutory solutions and their likely effects.

Special studies can play a particularly important role in this regard. In the early 2000s, RAND's *Aligning the Stars* report was used in support of the argument to lengthen military careers in order to capitalize more fully on the experience of long-term personnel (Harrell et al., 2004). As is often the case, this study utilized several databases and models to conduct its analysis. The RAND study utilized three databases to generate overall historical patterns, provide detailed information on common sequences of jobs, and generate inputs for modeling: the General and Flag Officer (G/FO) database maintained by the Directorate of Information Operations and Reports (DIOR),[17] the G/FO database maintained by DMDC,[18] and the Joint Duty Assignment Management Information System (JDAMIS).[19] The analysis was then conducted using two independent models (a steady-state system dynamic model and a more detailed entity-based model) and was subsequently validated on a third model to demonstrate that keeping very senior officers in service longer would not erode promotion opportunity

[17] The G/FO database maintained by DIOR is an aggregation of the General and Flag Officer Roster, an exhaustive list published monthly by DIOR. It tracks all active and reserve General and Flag Officers, maintaining information such as rank, specialty, service, job title, and unit.

[18] The G/FO database maintained by DMDC contains the personnel history (starting in September 1975) of all officers promoted to the rank of O-7 on or after January 1, 1990. Fields include name, date of birth, service, rank, occupational code, unit identification code, and unit address. The RAND report notes that the data in the DMDC database were more complete than those in the DIOR database but lacked a job title and unit name, which made it infeasible to use the DMDC data to perform the filtering of positions at the center of RAND's analysis.

[19] JDAMIS is a relational database containing data on "joint" positions and the officers who have served in at least one of those positions.

for the next cohort, provided prompt decisions were made about which officers truly merited that opportunity. While in its later action Congress left in place the looser age limits needed to implement change, it revoked some of the better pension rights awarded the most senior officers. The RAND report helped put to rest the fear that such a policy change would clog the promotion system. While other controversies made it difficult to secure and implement change, DoD did succeed in some loosening of the statutory restrictions.

Good data are especially important under a Declaration of National Emergency, which is how DoD currently operates, because many of the statutory restrictions on military personnel management (e.g., limits on the number of personnel in the more senior grades) can be waived in that circumstance. The personnel master files contain information on personal characteristics, such as name, social security number, date of birth, gender, race, ethnic group, and education, as well as information on military characteristics such as service, pay grade, months of service, and duty occupation (GAO, 2005). These data on servicemembers, which have been actively collected since 1971, give P&R the ability to judge whether requests for waivers are justified.

A different sort of waiver, likewise linked to career shaping, involves the statutory requirement that promotion to general or flag officer (i.e., one star and above) requires a certain amount of "joint" experience (essentially, experience in a non-Service-specific position, with the Joint Staff or a combatant command) (DoD, 2014). That experience is routinely credited when the billet is designated, but an issue arises when the content might meet the spirit of the statute even if the billet lacks such a designation. P&R is empowered to approve such a substitution and must judge based on descriptive material submitted by the military Service.

The issue of career shape is a central element of the current Secretary of Defense's Force of the Future initiatives, described in Chapter 2. Analogous issues related to career shaping and talent management arise in the private sector. They include recruiting, training, retaining, managing, promoting, compensating, and developing critical skills and talent across different parts of an enterprise (such as the military and civil service sectors of DoD). Other analogous issues include the sourcing of demand for DoD projects and tasks via servicemembers, civil service employees, and alternative options, as well as the allocation of human capital resources. Lessons learned and the combinations of predictive (statistical) analytics and prescriptive (optimization) analytics developed in these areas, some of which have been described above and others of which are described in Appendix D, may be useful to help address P&R requirements with respect to career shaping and talent management.

Ensuring That Programs Supporting DoD's Personnel Are Properly Structured and Provided

P&R is responsible for ensuring that various services supporting DoD personnel are properly structured and provided. Health care for military personnel and their families, and for military retirees and their families, is the leading example. Others include the availability of household goods, the education of military children overseas, and the opportunity for healthy leisure activity (mostly through nonappropriated fund activities). This responsibility is reflected not only in the Assistant Secretary for Health Affairs reporting to the Under Secretary, but also in an entire office focused on "Military Community and Family Policy."

The responsibility for providing key services, especially health care, drives some of P&R's most extensive data collection and analyses. It is responsible for the data systems that record all inpatient and outpatient care delivered in military facilities, as well as care purchased from private providers, including civilian hospital and nonhospital care to eligible dependents and retirees (through the Civilian Health and Medical Program of the Uniformed Services, commonly known as TRICARE) (Jansen, 2014). P&R is interested in the health status of its personnel and the productivity of its facilities—for instance, if vaccinations are up-to-date and how its clinicians compare to civilian practitioners. DoD is now developing a revised electronic health record for the care it delivers, which must be sharable with the Department of Veterans Affairs (DoD, 2015c).

One of the significant sources of health data originated from concerns that arose in the first Persian Gulf War about the effects of deployment on long-term health outcomes ("Gulf War Syndrome"). To improve its ability to assess the health effects of future exposures, DoD initiated the Millennium Cohort Study, where individuals are followed longitudinally and asked questions about post-traumatic stress disorder (PTSD) and marital status, for example. Scholars have used the data from the survey to examine questions related to health, sexual harassment, and employment (NRC, 2014).

Data sets assembled by P&R in the course of managing services extend well beyond health issues. For example, because it is responsible for MEPS, P&R oversees the administration of ASVAB and maintains the test results. In similar fashion, P&R is responsible for operating the schools for military children overseas (and, for historical reasons, in certain parts of the United States) through its DoD Educational Activity, which uses standardized test scores, such as TerraNova Assessment tests, to monitor the progress of students.

Anticipating and Responding to Sensitive Behavioral Issues

P&R must both anticipate and respond to sensitive behavioral issues. These include long-term issues such as the incidence of tobacco use and abuse of alcohol, and urgent issues such as sexual assault or constraints on religious expression. P&R's responsibility for dealing with issues arising from personal behavior engenders a number of specialized databases assembled both to gauge the extent of a problem and to monitor progress of policy actions taken. For example, the concern with sexual assault led P&R to conduct four major surveys of active duty troops, in 2002, 2006, 2010, and 2012; reserve components were surveyed similarly in 2004, 2008, and 2012; and DoD commissioned an independent study of the frequency and magnitude of unwanted sexual contacts (RAND, 2014).

P&R's survey instruments are designed to provide immediate answers, often by tabulating the frequency of responses after weighting the data to reflect the underlying population. A number of key questions are asked consistently over time so that trend analyses can be constructed (e.g., for satisfaction with military life, family satisfaction, intentions to remain in military service, stress). The instruments may also be used to estimate variables needed for specific policy purposes (e.g., estimates of family income to gauge the effect of policy changes). But as is the case with the sexual harassment/sexual assault surveys, the instruments are also designed to facilitate deeper analyses over a longer period of time. Typically, predictive models are constructed to explain the observed results as a function of (1) respondent background and (2) respondent behavior hypothesized to influence the observed results. These, in turn, are often employed for prescriptive purposes (e.g., to redesign elements of the compensation package).

REFERENCES

Baiocchi, D. 2013. *Measuring Army Deployments to Iraq and Afghanistan*. RAND Corporation. http://www.rand.org/content/dam/rand/pubs/research_reports/RR100/RR145/RAND_RR145.pdf.

Bartholomew, D.J. 1973. *Stochastic Models for Social Processes*. 2nd ed. London: Wiley.

Cao, H., J. Hu, C. Jiang, T. Kumar, T.-H. Li, Y. Liu, Y. Lu, S. Mahatma, A. Mojsilovic, M. Sharma, M.S. Squillante, and Y. Yu. 2011. OnTheMark: Integrated stochastic resource planning of human capital supply chains. *Interfaces* 41(5):414-435.

CRS (Congressional Research Service). 2012. *Troop Levels in the Afghan and Iraq Wars, FY2001-FY2012: Costs and Other Potential Issues*. Washington, D.C.

Defense Security Service. 2016. "Joint Personnel Adjudication System (JPAS)." http://www.dss.mil/diss/jpas/jpas.html. Accessed March 31, 2016.

Department of the Navy, Chief Information Officer. 2005. "The Defense Biometrics Identification System." *CHIPS*. October-December. http://www.doncio.navy.mil/chips/ArticleDetails.aspx?ID=3181.

DHRA (Defense Human Resource Activity). 2009. "Fiscal Year 2010 Budget Estimates." May. http://comptroller.defense.gov/Portals/45/Documents/defbudget/fy2010/budget_justification/pdfs/01_Operation_and_Maintenance/O_M_VOL_1_PARTS/DHRA.pdf.

Dietrich, B.L., E.C. Plachy, and M.F. Norton. 2014. *Analytics Across the Enterprise: How IBM Realizes Business Value from Big Data and Analytics*. Indianapolis: IBM Press.

DMDC (Defense Manpower Data Center). 2014. "The Strategic Planning Process at DMDC." https://www.dmdc.osd.mil/appj/dwp/documents.jsp.

DMDC. 2016a. "DMDC Overview." https://www.dmdc.osd.mil/appj/dwp/dmdc_overview.jsp. Accessed March 31, 2016.

DMDC. 2016b. "Personnel Data." https://www.dmdc.osd.mil/appj/dwp/personnel_data.jsp. Accessed March 31, 2016.

DoD (Department of Defense). 2011. "Department of Defense Instruction: Reserve Component Common Personnel Data System (RCCPDS)." Number 7730.54. Issued May 20. Under Secretary of Defense (Personnel & Readiness). http://www.dtic.mil/whs/directives/corres/pdf/773054p.pdf.

DoD. 2014. "Department of Defense Instruction: DoD Joint Officer Management (JOM) Program." Number 1300.9. Issued March 4. http://www.dtic.mil/whs/directives/corres/pdf/130019p.pdf.

DoD. 2015a. *2014 Demographics: Profile of the Military Community.* http://download.militaryonesource.mil/12038/MOS/Reports/2014-Demographics-Report.pdf.

DoD. 2015b. *Department of Defense Annual Report on Sexual Assault in the Military*. Washington, D.C.: Department of Defense.

DoD. 2015c. "DoD Awards Contract for Electronic Health Records." U.S. Department of Defense News article. July 29. http://www.defense.gov/News-Article-View/Article/612714.

DoD. 2016. *Population Representation in the Military Services, Fiscal Year 2014*. Office of the Under Secretary of Defense (Personnel & Readiness). https://www.cna.org/pop-rep/2014/. Accessed February 15, 2016.

DPCLD (Defense Privacy and Civil Liberties Division). 2009. "Non-combatant Tracking System (NTS) & Evacuation Tracking and Accountability System." System of Record Notices (SORNs): Office of the Secretary of Defense and Joint Staff (OSD/JS): DMDC 04. http://dpcld.defense.gov/Privacy/SORNsIndex/DOD-Component-Notices/OSDJS-Article-List/.

DTIC (Defense Technical Information Center). 2014. "Department of Defense Instruction: Defense Incident-Based Reporting System (DIBRS)." Number R 7730.47. January 23. http://www.dtic.mil/whs/directives/corres/pdf/773047p.pdf.

Gael, S. 1988. *Job Analysis Handbook for Business, Industry and Government.* New York: Wiley.

GAO (Government Accountability Office). 2005. *Military Personnel: Reporting Additional Servicemember Demographics Could Enhance Congressional Oversight: Report to Congressional Requesters*. GAO-05-952. Washington, D.C.

Gao, X., Y. Lu, M. Sharma, M.S. Squillante, and J.W. Bosman. 2013. Stochastic optimal control for a general class of dynamic resource allocation problems. Pp. 3-14 in *Proceedings of the IFIP WG 7.3 Performance Conference*. Vienna, Austria, September 24-26.

Harrell, M.C., H.J. Thie, P. Schirmer, and K. Brancato. 2004. *Aligning the Stars: Improvements to General and Flag Officer Management*. Santa Monica, Calif.: RAND Corporation. http://www.rand.org/pubs/monograph_reports/MR1712.

Hoffmann, C., E. Lesser, and T. Ringo. 2012. *Calculating Success: How the New Workplace Analytics Will Revitalize Your Organization*. Boston: Harvard Business Review Press.

Hu, J., B.K. Ray, and M. Singh. 2007. Statistical methods for automated generation of service engagement staffing plans. *IBM Journal of Research and Development* 51(3):281-293.

Jansen, D.J. 2014. *Military Medical Care: Questions and Answers*. Congressional Research Service. https://www.fas.org/sgp/crs/misc/RL33537.pdf.

McCloy, R.A., D.A. Harris, J.D. Barnes, P.F. Hogan, D.A. Smith, D. Clifton, and M. Sola. 1992. *Accession Quality, Job Performance, and Cost: A Cost/Performance Trade-off Model*. FR-PRD-92-11. Alexandria, Va.: Human Resources Research Organization.

NRC (National Research Council). 2014. *The Context of Military Environments: An Agenda for Basic Research on Social and Organizational Factors Relevant to Small Units*. Washington, D.C.: The National Academies Press.

OMB (Office of Management and Budget). 2003. "Circular No. A-76 Revised." May 29. https://www.whitehouse.gov/omb/circulars_a076_a76_incl_tech_correction/.

PERSEREC (Defense Personnel and Security Research Center). 2016. "PERSEREC Initiatives." http://www.dhra.mil/perserec/currentinitiatives.html. Accessed March 31, 2016.

Putka, D.J. 2004. "An Evaluation of Moral Character Waiver Policy: Database Documentation." FR-04-17. Alexandria, Va.: Human Resources Research Organization.

Putka, D.J., C.L. Noble, D.E. Becker, and P.F. Ramsberger. 2003. "Evaluating Moral Character Waiver Policy Against Servicemember Attrition and In-service Deviance Through the First 18 Months of Service." FR-03-96. Alexandria, Va.: Human Resources Research Organization.

RAND. 2014. *The RAND Military Workplace Study: Sexual Assault and Sexual Harassment in the U.S. Military*. Santa Monica, Calif.: RAND Corporation.

Tillson, J.C.F., R.J. Atwell, J.R. Brinkerhoff, W.R. Burns, Jr., M. Burski, J. Castillo, M. Diascro, R. Fabrie, W.D. Freeman, M.R. Lewis, C. Lyman, and L. Morton. 2000. *Independent Review of DoD's Readiness Reporting System*. IDA P-3569. Alexandria, Va.: Institute for Defense Analyses.

U.S. Army. 2011. "Defense Readiness Reporting System-Army Procedures: Commander's Unit Status Reporting Procedures." Army Pamphlet 220-1. November 16. http://ssitoday.armylive.dodlive.mil/files/2014/03/DA-PAM-220_1.pdf.

U.S. Army. 2012. "Fully Automated System for Classification (FASCLASS)." http://cpol.army.mil/library/permiss/38.html.

Vajda, S. 1978. *Mathematics of Manpower Planning*. New York: Wiley.

Vie, L.L, K.N. Griffith, L.M. Scheier, P.B. Lester, and M.E.P. Seligman. 2013. The Person-Event Data Environment: Leveraging big data for studies of psychological strengths in soldiers. *Frontiers in Psychology* 4:934.

Vine, D. 2015. Where in the World Is the U.S. Military? *Politico Magazine*, July/August. http://www.politico.com/magazine/story/2015/06/us-military-bases-around-the-world-119321.

White, H.C. 1970. *Chains of Opportunity: System Models of Mobility in Organizations*. Boston: Harvard University Press.

4

Overview of Data Science Methods

INTRODUCTION

Data applicable to personnel and readiness decisions are increasing rapidly as is the potential to make meaningful decisions enhanced by previously inaccessible information. Automation of tracking, the increase of new data types (e.g., social media, audio, video), enhanced storage of electronic records, repurposing of administrative records, and the explosion of modeling data have all increased the availability of data. However, making full use of these data requires not only proper storage and management (Dasu and Johnson, 2003; Wickham, 2014) but also advanced analytical capabilities.

This report considers data science in its broadest sense, as a multidisciplinary field that deals with technologies, processes, and systems to extract knowledge and insight from data and supports reasoning and decision making under various sources of uncertainty. Here, two aspects of data science are of interest: (1) the management and processing of data and (2) the analytical methods and theories for descriptive and predictive analysis and for prescriptive analysis and optimization. The first aspect involves data systems and their preparation, including databases and warehousing, data cleaning and engineering, and data monitoring, reporting, and visualization. The second aspect involves data analytics and includes data mining, text analytics, machine and statistical learning, probability theory, mathematical optimization, and visualization.

This chapter discusses some of the data science methods and practices being employed in various domains that pertain to the analysis capabilities relating to personnel and readiness missions in the Department of Defense

(DoD). Many of these methods are well known to the personnel and readiness community. Some of the important considerations discussed include how descriptive and predictive analytics methods can be used to better understand what the data indicate; how decision making under uncertainty can be enhanced through prescriptive analytics; and the usefulness and limitations of these approaches.

The proliferation of data and technical advances in data science methods have created tremendous opportunities for improving these analysis capabilities. This section focuses on data science methods, including those associated with data preparation and descriptive, predictive, and prescriptive analytics, thus providing some of the technical details and foundation for the data science methods that will be referenced in subsequent chapters. Figure 4.1 illustrates the typical evolution from data sources to analysis results. This chapter does not discuss considerations needed for all modeling (e.g., to avoid overfitting data and to evaluate stability in terms of cross-validation) but instead offers a nonexhaustive list of methods to introduce some key approaches relevant to the missions of the Office of the Under Secretary of Defense (Personnel & Readiness), referred to throughout the report as P&R. Many of these methods are well known to researchers in the personnel and readiness community and are only cursorily discussed, while others that the committee believes are not widely used have been explained in more detail. This chapter discusses just the methods, while Chapter 7 provides some examples of how the methods could be applied to P&R mission areas.

The increase in the volume of data does not in and of itself lead to better outcomes. There are challenges associated with storing, indexing, linking, and querying large databases, but perhaps the most significant challenge is drawing meaningful inferences and decisions from analysis of the data. As discussed in the 2013 National Research Council report *Frontiers in Massive Data Analysis*, "Inference is the problem of turning data into knowledge, where knowledge often is expressed in terms of entities that are not present in the data per se but are present in models that one uses to interpret the data" (NRC, 2013, p. 3). In drawing a meaningful inference, issues such as sampling bias (where a sample is collected in such a way that some members of the intended population are less likely to be included than others), provenance (where layers of inference are compounded, obscuring the original data), and compounding error rates (where several hypotheses are considered at the same time) all play a role in achieving quality results (NRC, 2013). This of course assumes that statistical and decision-making techniques are being applied properly and that the results are reproducible (NASEM, 2016).

The following section on data preparation discusses common data preparation tasks and methods, while the subsequent section on data analytics discusses descriptive and exploratory analysis, predictive analysis, and prescriptive analysis.

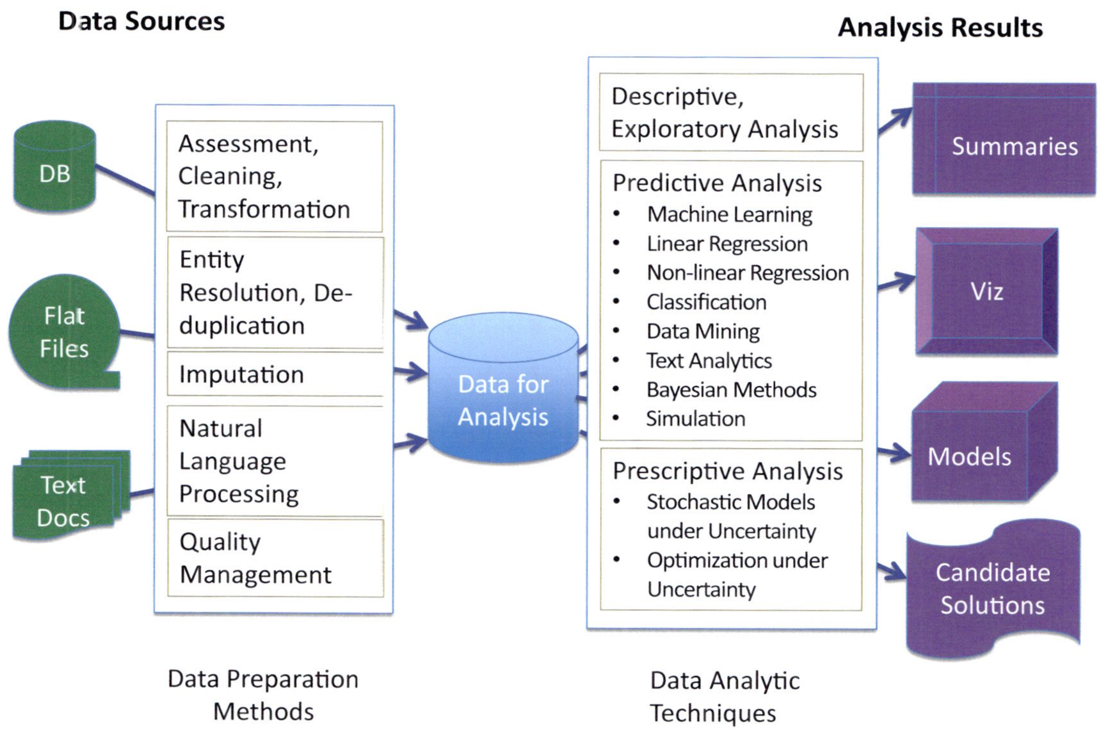

FIGURE 4.1 The evolution from data sources to analysis results passes through several steps. Raw data (captured in databases [DB], flat files, and text documents) must first go through various data preparation methods to prepare them for analysis. The prepared data can then be analyzed using a variety of data analytic techniques to summarize and visualize the data and develop models and candidate solutions.

DATA PREPARATION

The foundation of data analytics is having appropriate data that are of sufficient quality, appropriately organized for the analysis task at hand. Unfortunately, data seldom arrive in a suitable state, and a necessary first step is getting them into a proper form to support analysis. This step, sometimes referred to as "data wrangling," has been estimated to occupy up to 80 percent of the total analysis time according to a recent survey of data scientists by CrowdFlower (Biewald, 2015). This intensive overhead cuts down on the effective productivity of data analysts and may deter them from undertaking certain studies. The following sections discuss some of the tasks involved in preparing data for analysis and some techniques to assist analysts with those tasks.

Common Data Preparation Tasks

Data in their original form are not typically ready to be used without some initial work. This section discusses some steps commonly taken to prepare data for analysis, including locating, acquiring, and ingesting data; assessing and cleaning data; reconciling and making data uniform; extracting, restructuring, and linking data; coding and annotating data; and updating data as new information becomes available. These issues are discussed in this section, and ways to approach them will be discussed in the following section.

Data Location, Acquisition, and Ingestion

An analyst might need to first determine whether data suitable for a particular analysis exist in a particular organization and, if they do exist, how they can be accessed or how a copy can be obtained. Once the data are in hand, the analyst might need to load them onto a different system to work with them. If the data need to be loaded into a database management system (DBMS), then the analyst or database administrator will need to provide a database schema (definitions of the different tables and their formats) and possibly write scripts to ingest the data into the DBMS.

Data Assessment and Cleaning

Data from transactional (administrative) systems are often repurposed for analysis projects. Data that might be sufficient for operational use might nevertheless present problems for analysis. For example, if a "language skills" field in a personnel record is examined only by humans, then variation in values such as "Chinese (Cantonese)," "Cantonese," or "Chinese

(Yue dialect)" might present few analytical challenges to a human with some knowledge of linguistics. However, if examined by a computer algorithm, it might have to be regularized before use with data analysis routines. Administrative data can also have spurious and missing values, differences in data representations (such as different formats for dates and times), compound values (e.g., a list of languages instead of a single language), and various kinds of noise (such as those that arise from data-entry errors). An analyst wanting to repurpose a data source will need to spend time assessing what kinds of quality problems might exist in the data and whether the data can be adapted for the current study. Once quality is assessed, there will likely be a data-cleaning phase, in which errors, inconsistencies, and duplicates are detected and either corrected or excluded. Even when data have been specifically collected for a study—rather than being repurposed from another use—there can be quality problems. For example, surveys can have a missing or inappropriate response or have entire sections that were not completed or failed to cover certain segments of the target population.

Data Uniformity and Reconciliation

If an analysis is to use multiple data sources, how a particular data item is interpreted or collected across sources can vary. For example, one source might break down spoken versus written language skills, while another lists only the language. Or, one source might refer to explicit instruction or testing in a language, while another relies on self-reported capabilities. Reconciling such variation will ultimately require the analyst to decide what is appropriate for a given study. However, there is sometimes little or no clear documentation on the precise meaning of different fields, complicating the job of the analyst.

Data Extraction, Restructuring, and Linking

Data often arrive in a form that does not directly match the input requirements of the analysis tools, especially if they are being repurposed from another use. Some data values might be captured in free-text fields, a data set as a whole might have the wrong organization, or the values for analysis could be split up across several files. As an example of wrong organization, the analyst might have a collection of records, each with a date of enlistment and current pay rate. A statistical package that would correlate the two might want two parallel lists, one with enlistment dates and the other with pay rates. Overcoming such problems can necessitate a significant effort by the analyst to pull appropriate values from text fields (or write a script that will do this), to restructure the data, or to link corresponding records across two or more files.

Data Coding and Annotation

Sometimes in preparation for analysis, data values need to be mapped to a fixed set of categories, particularly for survey responses. For example, a field listing a job title might need to be first classified as to "Service," "manufacturing," "construction," and so forth. Data might need to be further "adorned" with additional information before analysis. For example, a geographic analysis might require addresses in the data to be "geocoded": translated in latitude–longitude pairs.

Data Currency and Refresh

Most administrative data sources are not static. It may take considerable time to acquire a data set, ingest it, clean it, and otherwise prepare it for analysis, by which time the data may be out-of-date. Even if existing records do not change (e.g., history of past pay slips), a data source may be augmented periodically (i.e., new pay slip records added every 2 weeks), meaning a copy of that source will need to be refreshed if it is to remain current. It might seem that using a data source in situ at its original location would avoid problems with stale or incomplete data. However, such use is often not feasible because the data source cannot be accessed remotely, the host system does not want to allow the additional processing load of analytic queries, or the data cannot be cleaned in place.

Data Preparation Methods

With such large amounts of a data analyst's time being spent on data preparation, techniques that reduce such overhead have high value. This section describes some of the main data preparation methods.

Reusing Attention

Avoiding data preparation tasks by capturing results of previous work is the first step in reducing an analyst's load. The bounding constraint in data analysis is often human attention, so reusing that attention where possible helps conserve that resource. Too often, data preparation work results in a spreadsheet or data set that resides on a personal workstation and is not even visible to others who might benefit from its use. Having shared repositories of data with various degrees of preparation can at least spread the investment of human attention. In commercial practice, such repositories span the range from "data lakes" (Stein and Morrison, 2014) that merely collect raw data in one place to full-fledged data warehouses[1] that regularize, clean, and integrate data according to a common schema.

Some authors have suggested a virtual warehouse approach that leaves data in their source system and provides a federated search or query capability over the collection of sources. This approach does not seem good for P&R analyses for several reasons:

1. The original data often reside in production systems that would not necessarily tolerate the additional load of analytic queries.
2. Federated approaches are predicated on using a common query language, such as SQL, across sources that share a data model, such as relational databases, a requirement that is not met by many of the potential data sources the committee considered.
3. The federated approach generally does not accommodate transformed or restructured versions of the data, as it is usually not possible to create arbitrary new data sets at the sources.
4. Federation entails repeated transfer of information to answer various requests, with each transfer increasing the risk of data interception, either in transit or at the originating or receiving system. This risk may not be tolerable for many of the sensitive and confidential sources in this domain.

Data Assessment, Cleaning, and Transformation

For structured data, such as relational databases,[2] there is a wide range of mature commercial tools for data preparation, especially in connection with data warehousing and data integration activities (Rahm and Do, 2000). Data profiling tools have been available for decades but are still the subject of active research (Naumann, 2013). Data profiling collects statistics and other information about a data set, such as min and max values, frequent values, and outliers, in order to understand the nature and quality of the data before further processing. Extract-transform-load (ETL) tools have been around since the advent of data warehousing (Kimball and Caserta, 2004). Such tools help extract data from source systems, transform it appropriately, and then load it into a target system, often a data warehouse. The transform stage is generally the richest, having a rule-driven framework encompassing both data manipulation and data validation.

---

[1] A data warehouse can be viewed as a kind of database, organized to facilitate reporting and analysis. Hence it often combines data from multiple sources across an enterprise and is generally updated periodically in a batch fashion, in contrast to online transaction processing (OLTP) databases, organized to support small, frequent updates.

[2] Relational databases are databases structured to recognize relations among stored items of information—for example, by storing data in tables with rows and columns.

Data manipulation includes dropping columns, recoding values, calculating new columns, splitting or combining values, joining tables, and aggregating or reordering data. Data validation can include checking format, testing values ranges, and look-up in tables of legal values.

More recently, tools have appeared to work with broader classes of data. One example is OpenRefine (formerly Google Refine), an open-source tool for data cleaning and transformation that can work with CSV files, XML data, RDF triples, JSON structures, and other formats (Verborgh and De Wilde, 2013). The PADS project (Fisher and Walker, 2011) works with an even broader class of inputs, so-called ad hoc data formats. Ad hoc data formats are those arising from particular applications where there is no existing base of tools for manipulating the data and can arise in areas such as telecommunications, health care, sensing, and transportation. PADS uses a data description of a given source to generate a range of tools, such as parsers, validators, statistical analyzers, and format converters. Furthermore, PADS facilitates learning the data description given a set of examples from a source.

Entity Resolution and De-Duplication

A common task in data preparation is identifying multiple records that refer to the same thing. This task can arise for many reasons, such as the lack of a clean database (e.g., an address list that has repeated information) or combining two data sources about the same subject (even if the individual sources are duplicate-free). There is a large body of tools to handle this problem, known variously as entity resolution, object identification, reference reconciliation, and several others (Getoor and Machanavajjhala, 2012). These tools use a variety of approaches, such as approximate match, clustering, normalization, probabilistic methods, and even crowdsourcing.

Data Imputation

Missing values can cause problems for various kinds of analytic methods. In the face of such problems, one can seek alternative methods that tolerate missing data, or one can impute the missing value (fill it in with an estimate). Imputation methods can be as simple as inserting a default value or more complex, such as consulting a historical archive or using a statistical model. However, not all imputation methods are suitable for applying at the data preparation stage; rather, they are applied as part of analysis. For example, multiple imputation constructs several data sets from an initial data set with missing values, then runs the analysis on each and combines the results (Enders, 2010).

Natural Language Processing

Multiple individuals[3] in both the government and commercial sectors told the committee that natural language processing (NLP) would soon "be ready for prime time," for use in data preparation tasks. For example, the committee heard that the Defense Manpower Data Center's (DMDC's) Data Science Program is testing NLP methods against human coding of free-text data as part of a task to find predictors of outcomes for appeals of security clearance decisions. There are already some specialized tasks where NLP approaches have been very successful, such as named entity recognition (NER). NER is the process of extracting phrases from text that identify specific entities such as persons, places, and organizations. NER techniques are quite robust, and some work across multiple languages (Al-Rfou et al., 2015). Much current research focuses on broader NLP tasks, such as text analytics, which seeks to extract text features that can be used with structured analysis methods. Text analytics supports applications such as sentiment analysis, relationship extraction, and medical-record coding.

Automated Data-Quality Management

The advent of big data in many domains means that manual methods of data-quality assurance are no longer feasible. Thus, automated data-monitoring techniques are of increasing interest, with much of the initial work carried out in the context of data coming from sensors. Early efforts rely on human construction of rules that apply "sanity checks" to the data. For example, the National Oceanic and Atmospheric Administration's guide for automated checking of buoy data has a rule that checks that near-shore instruments do not report significant swells originating from the direction of land (NDBC, 2009). However, there are machine-learning approaches that require less human intervention (Isaac and Lynes, 2003; Smith et al., 2012).

DATA ANALYTICS

Although some scientists and institutions have attempted to define the broad area of data analytics, it is difficult to do so. Davenport and Harris (2007) define data analytics to be the "extensive use of data, statistical and quantitative analysis, explanatory and predictive models, and fact-based management to drive decisions and actions." Using the often-cited diagram in Figure 4.2, they go on to characterize analytics as starting with statisti-

---

[3] These discussions occurred during the committee's meetings, site visits, and follow-up discussions with relevant individuals. Please see Appendix C for a full list of public committee meetings and the Acknowledgments section of the front matter for a list of individuals who provided input to the study.

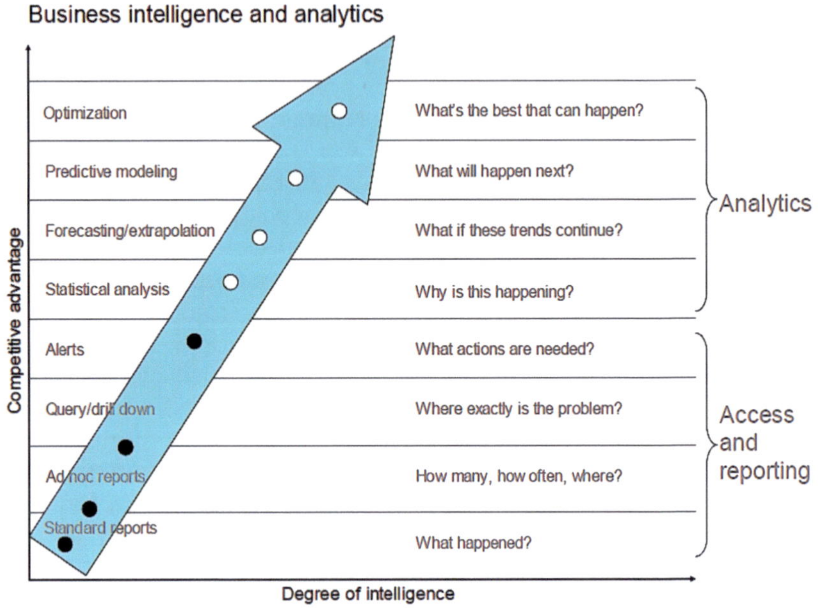

FIGURE 4.2 Business intelligence and analytics. SOURCE: Davenport and Harris (2007).

cal analysis ("Why is this happening?"), then moving on to forecasting/extrapolation ("What if these trends continue?"), then moving to predictive modeling ("What will happen next"), and ending with optimization ("What is the best that can happen?"). Even before the book by Davenport and Harris, the term "predictive analytics" was used as early as 1999 (Brenda Dietrich, IBM, personal communication, September 23, 2015), and as early as 2005 IBM was using the term "prescriptive analytics" (Brenda Dietrich, IBM, personal communication, September 23, 2015), which corresponds to what Davenport and Harris called optimization.

Analytics is defined by Lustig et al. (2010) to comprise descriptive, predictive, and prescriptive analytics, where descriptive analytics is defined as a set of technologies and processes that use data to understand and analyze an organization's performance; predictive analytics is defined as the extensive use of data and mathematical techniques to uncover explanatory and predictive models for an organization's performance as represented by the inherent relationship between data inputs and outputs/outcomes; and prescriptive analytics is defined as a set of mathematical techniques that computationally determine a set of high-value alternative actions or decisions given a complex set of objectives, requirements, and constraints, with the goal of improving organizational performance. In their recent book, Dietrich et al. (2014) characterize descriptive analytics as reporting what has happened, analyzing contributing data to determine why it happened, and monitoring new data to determine what is happening now; predictive analytics as using techniques such as statistics and data mining to analyze current and historical information to make predictions about what will happen in the future, typically producing both a statement of possible events that could occur and the associated probabilities of their occurring; and prescriptive analytics as covering analytics methods that recommend actions with the goal of finding an action (or set of actions) that will maximize the expected value (e.g., utility) associated with the outcome.

Models are at the heart of many analytical approaches, from simple descriptive summaries to detailed theoretical representations grounded in physical laws. Statistical models are a central component of all analytics. Data (observations) are integrated through statistical representations to create empirical understandings of reality. Descriptive analytics can characterize the underlying population represented by the data, and the processes that generated the data, in order to make predictive inferences about the population. Probability models are at the foundation of statistical models that enable the descriptions of the observed data to formulate inferences about the population from which these data were observed.

Leek and Peng (2015) provide a helpful overview of models and the types of questions they can address. They present six types of models from descriptive to causal (see Figure 4.3). This overview is a helpful characterization of moving from raw data to predictive analytics. They also highlight the common mistakes in the use of models such as using correlation to imply causation or using exploratory analyses to imply a predictive model structure.

In 2010 the Institute for Operations Research and the Management Sciences (INFORMS) defined analytics as comprising descriptive, predictive, and prescriptive analytics, focusing on predictive analytics based on methods from data mining, machine learning, and statistics, and on prescriptive analytics based on methods from stochastic modeling and mathematical optimization.[4] Following in kind, this section presents data analytics methods in those same terms.

As described in detail above, data cleaning and linking challenges need to be overcome before any analysis can occur. In most cases, study resources are largely utilized during these stages. Predictive and prescriptive models require follow-on efforts to calibrate, correct, and reanalyze results, which makes them difficult to undertake in a resource-constrained environment

[4] The website for the Analytics Society of INFORMS is https://www.informs.org/Community/Analytics, accessed October 7, 2016.

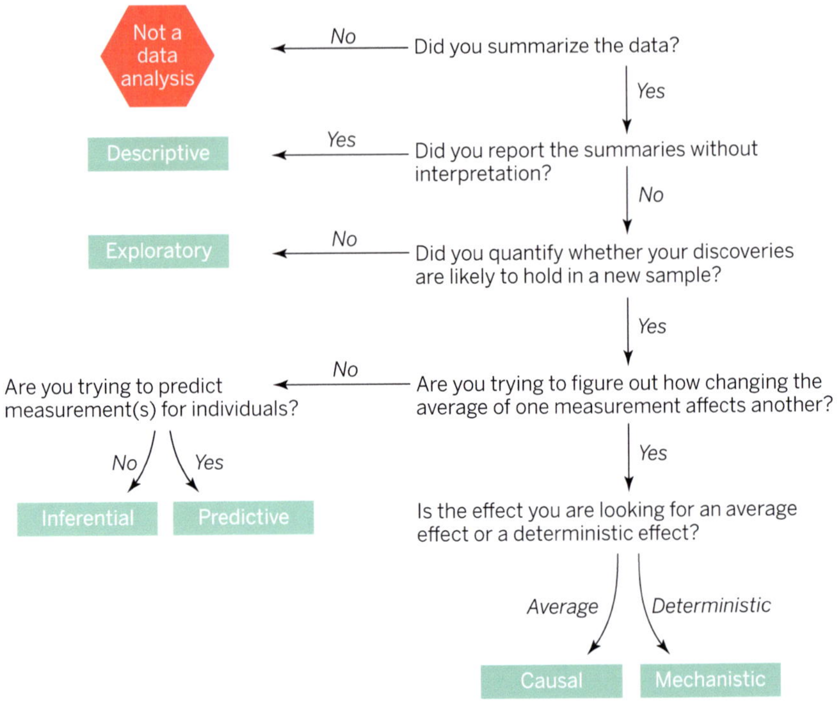

FIGURE 4.3 Data analysis flowchart. SOURCE: Leek and Peng (2015).

where urgent decisions for resolving pressing problems are needed. Lastly, the confidence and significance of results need to be assessed for all data analyses. It is important to validate all results to see if they make sense and represent effects as expected.

The following sections describe some approaches toward descriptive and exploratory analysis (using data to summarize and visualize the current state), predictive analysis (using data to determine patterns and predict future outcomes and trends with methods such as linear and nonlinear regression, classification, data mining, machine learning, text analysis, Bayesian methods, and simulation), and prescriptive analysis (using data to determine a set of decisions and/or actions that gives rise to the best possible results based on various predicted outcomes from predictive analytics and subject to various constraints using stochastic models of uncertainty, mathematical optimization under uncertainty, and optimal solutions).

Descriptive and Exploratory Analysis

Descriptive analytics are the most common form of data analytics because their primary purpose is to summarize and understand existing data. They are typically the least challenging because they seek to summarize measurements in a data set without further interpretation. Exploratory data analysis goes a step further and builds on a descriptive analysis by searching for discoveries, trends, correlations, or relationships between the measurements to generate ideas or hypotheses (Leek and Peng, 2015).

The data used in descriptive and exploratory data analysis can be defined as categorical/discrete or continuous. Discrete data consist of particular finite or countably infinite values in a given discrete data set and can be numerical (e.g., the number of personnel deployed) or categorical (e.g., male or female). Continuous data, on the other hand, result when an observation can take on any value within a certain range or interval (e.g., temperature readings).

These data can be summarized in a number of ways. Frequencies, distributions, and tabulations are used to examine the count of the occurrences of values within a particular group or interval. Calculating the central tendency—the mean, median, and mode of the data—summarizes the data into a single value that is typical or representative of all the values in the data set. Assessing the spread—the range, quartiles, variance, or standard deviation—shows how scattered the values are and how much they differ from the mean value.

The visualization or presentation of these data and analyses is an important means of conveying information. While there are a variety of ways to do this in a pictorial or graphic format, visualizations using bar charts, box plots, and scatter plots are common approaches.

Predictive Analysis

Predictive analysis goes a step beyond descriptive and exploratory analysis by extracting information from data sets to determine patterns and predict future outcomes and trends. Predictive analytics can be targeted to test a particular hypothesis or exploratory to formulate hypotheses (Hastie et al., 2008; NRC, 2013). There are a number of tools used for these analyses, some of the most common of which are briefly described in the following subsections.

Machine Learning

Machine learning methods were developed to deal with the need for out-of-sample prediction and address the problems of dealing with massive data sets containing many predictors. Machine learning develops techniques for teaching computers to act without explicitly programming them. These techniques fall into three broad classes:

- *Supervised.* A teacher provides the computer with explicit examples (say, of a concept being learned) or feedback on the correctness of a particular decision.
- *Unsupervised.* The computer seeks to uncover hidden patterns without explicit labeling of examples or an error signal.
- *Reinforcement.* A software agent determines how to optimize its behavior from a local reward signal, but without explicit input–output pairs or feedback on suboptimal actions.

A wide range of such algorithms exist (over 50 different supervised learning algorithms now in the statistical computing software R[5]), and these different algorithms work well in different settings. Several "ensemble methods" have been developed to combine predictions across a range of learning algorithms to arrive at optimal predictions. All these methods are designed to maximize prediction at the expense of allowing the analysts to evaluate the effects of individual predictors. However, these predictors are often highly intercorrelated, and models with highly intercorrelated predictors usually have poor out-of-sample performance.

Machine learning tasks include learning a model that can predict discrete categories (classification, discussed below) or a continuous output (regression, discussed below); dimensionality reduction to simplify a multidimensional data set; clustering input data into cohesive groups; multivariate querying to find the objects most similar to each other or a particular candidate; and density estimation of an unobservable underlying density function.

Linear Regression

Linear regression is a widely used approach to modeling the relationship between a dependent variable and one or more explanatory variables using linear predictor functions, where unknown model parameters are estimated from the data. A common problem of dealing with large data sets is that it is difficult to develop models with thousands of predictors. Linear regression has many practical uses, particularly for predicting, forecasting, reducing errors, and quantifying the strength of the relationship between data. This type of analysis is often used because models that depend linearly on their unknown parameters are easier to fit than models that are non-

[5] See the "R Project for Statistical Computing" website at https://www.r-project.org/.

linearly related to their parameters. Linear regression models are often fitted using the least squares approach, but they may also be fitted in other ways, such as by minimizing the "lack of fit" in some other mathematical norm (as with least absolute deviations regression), or by minimizing a penalized version of the least squares loss function as in ridge regression (L_2-norm penalty) and lasso (L_1-norm penalty) (Friedman et al., 2010).

Dimension reduction is often conducted to reduce the number of random variables under consideration. Two popular methods are the lasso and elastic nets. The lasso is a shrinkage and selection method for linear regression that minimizes the usual sum of squared errors, with a bound on the sum of the absolute values of the coefficients (Tibshirani, 1996). The lasso is a penalized least squares method imposing an L_1-penalty on the regression coefficients and simultaneously provides both continuous shrinkage and automatic variable selection. However, the lasso is subject to limitations in some cases, including limited or unspecific variable selection and prediction performance dominated by the underlying ridge regression (Tibshirani, 1996).

The elastic net is a regularization and variable selection method that generalizes and outperforms lasso in some cases. Elastic net encourages a grouping effect, where strongly correlated predictors tend to be in or out of the model together, and is particularly useful when the number of predictors is much bigger than the number of observations (Zou and Hastie, 2005).

Nonlinear Regression

Nonlinear regression is a form of regression analysis in which observational data are modeled by a function that is a nonlinear combination of the model parameters and depends on one or more independent variables. The data are fitted by a method of successive approximations. In contrast to linear analysis, where the linear equation has one basic form, nonlinear equations can take many different forms. Nonlinear regression is typically computationally intensive and requires an iterative approach to solve, in contrast to linear regression, in which many solutions can be calculated without advanced computation.

Kernel methods are a class of algorithms for pattern analysis of computational tools used for nonlinear regression analysis, including support vector machines (SVMs). SVMs are supervised machine learning models with associated learning algorithms that analyze data and recognize patterns. While SVM models are typically used for classification (Burges, 1998)—by representing example data as points in space, mapping example data into separate categories divided by a clear gap, mapping new data into that same space, and predicting which category data belong to based on which side of the gap they fall on—they can also be used for regression analysis (Vapnik, 1995). SVMs are largely characterized by the choice of their kernels, and SVMs thus link the problems they are designed for with a large body of existing work on kernel-based methods (Smola and Schölkoph, 1998).

Classification

Classification is the problem of identifying a set of categories to which a new observation belongs, on the basis of a training set of data containing observations whose category membership is known (e.g., differentiating e-mail messages to identify which should be filtered as spam). Classification is an instance of supervised machine learning where a training set of correctly identified observations is available. The corresponding unsupervised procedure is known as clustering and involves grouping data into categories based on some measure of inherent similarity or distance. Often, the individual observations are analyzed into a set of quantifiable properties, known variously as explanatory variables or features. These properties may variously be categorical (such as "A," "B," "AB," or "O," for blood type), ordinal (such as "large," "medium," or "small"), integer-valued (such as the number of occurrences of a particular word in an e-mail), or real-valued (such as a measurement of blood pressure). Other classifiers work by comparing observations to previous observations by means of a similarity or distance function. An algorithm that implements classification, especially in a concrete implementation, is known as a classifier. The term "classifier" sometimes also refers to the mathematical function, implemented by a classification algorithm, that maps input data to a category (Tang et al., 2014).

Classification uses categorical data and is often done with logistic regression or classification trees. Logistic regression measures the relationship between the categorical dependent variable and one or more independent variables by estimating probabilities using a logistic function, which is the cumulative logistic distribution (Cox, 1958). Classification trees are used to predict membership of cases or objects in the classes of a categorical dependent variable from their measurements on one or more predictor variables. The goal of classification trees is to predict or explain responses on a categorical dependent variable; as such, the available techniques have much in common with the techniques used in the more traditional methods of discriminant analysis, cluster analysis, nonparametric statistics, and nonlinear estimation. The flexibility of classification trees makes them a very attractive analysis option (Hill and Lewicki, 2006).

Regression tree analysis, in which a predicted outcome can be considered a real number, is also used in addition to classification tree analysis, in which the predicted outcome is the class to which the data belong. Classification and Regression Tree (CART) analysis is an umbrella term used to refer to both of the above procedures (Breiman et al., 1984).

Data Mining

Data mining (also called knowledge discovery) is the computational process of discerning a previously unknown pattern in a data set and transforming it into an understandable structure for further use. It lies at the intersection of statistics, machine learning (see below), and data management. While many data-mining techniques have been known for decades, the focus of late has been on methods that can work with large data sets (too large to fit in main memory). For example, while decision trees have been studied since the early 1980s, in the mid-1990s methods began to appear to build decision trees on disk-based data with limited passes over the data. Some of the main classes of data-mining algorithms are the following:

- *Cluster detection.* Trying to find natural groupings of data, for example, to help with summarization of a data set.
- *Anomaly detection.* Finding outliers in a data set, for example, to detect fraudulent banking transactions.
- *Association rule learning.* Discovering relationships between variables in data sets, for example, market-basket analysis, which tries to determine which items are commonly purchased together.

Sometimes the outputs of data-mining methods can be used directly to influence future action, or they might suggest hypotheses to test more rigorously with other methods (see Hastie et al., 2008).

Text Analytics

Text analytics (also called text data mining) seeks to extract useful, and generally machine-processable, information from unstructured or semi-structured textual sources. It brings to bear techniques from a variety of areas, such as natural language processing (part-of-speech tagging, topic modeling, co-reference determination), information extraction (named-entity recognition, relationship extraction), information retrieval (novelty detection), and statistics (pattern learning). The great increase of textual sources—such as social media, online product and business reviews, and transcribed call records—have piqued interest in automated text analysis methods as data volumes have outstripped the capacity of human analysts. Application areas include sentiment analysis for brands, root cause determination in customer complaints, employee sentiment, and identification of insurance claims for subrogation.

NLP techniques are reaching a level of maturity such that they are suitable for preparation and analysis of unstructured text data. For example, Intel has begun using NLP techniques to help in analyzing the tens of thousands of responses to its annual organizational health survey. Companies are also turning more to external sources of data, such as demographic and labor data from both public and private sources. One study revealed that the true supply of candidates in underrepresented groups was much lower than publicly reported, leading to a shift in diversity efforts from recruitment to retention. In contrast, Intel noted that social media data were becoming less useful for studying current employees, as users are becoming savvier about their privacy settings. However, internal media, such as e-mail and discussion boards, can have value. For example, simply analyzing the promptness of replies to e-mail can reveal much about intra- and intergroup dynamics. Some companies have gone as far as instrumenting employees to track the scope and frequency of interpersonal communication. Olguín-Olguín and Pentland (2010) report on several studies where sensors were used to gather data on face-to-face interactions (sometimes together with electronic communications). Analysis of the resulting graph structures revealed a correlation between site productivity and difference in job attitude.

Bayesian Methods

Bayesian methods are powerful for integrating multiple types and sources of data; they cross into all areas of predictive analysis because they represent a state of knowledge or a state of belief via a probability distribution. The key ingredients for a Bayesian analysis are the likelihood function, which reflects information about the parameters contained in the data, and the prior distribution, which quantifies what is known about the parameters before observing data. The prior distribution and likelihood can be easily combined to form the posterior distribution, which represents total knowledge about the parameters after the data have been observed. Simple summaries of this distribution can be used to isolate quantities of interest and ultimately to draw substantive conclusions (Glickman and van Dyk, 2007).

Simulation

Simulation, sometimes referred to as modeling, simulation, and analysis (MS&A), is a form of predictive analytics useful for scenario development and analysis and what-if studies. As with all forms of predictive analytics, it relies heavily on a descriptive understanding of the underlying population or phenomena of interest (reality being studied). DoD has a long history of using MS&A to study future scenarios, including force planning and war gaming. *Defense Modeling, Simulation, and Analysis: Meeting the Challenge* gives both a historical look back and a future outlook for MS&A in face of rapid changes within DoD that are affecting both composition of the force and future conflicts (NRC, 2006).

It is important to recognize that simulation answers the what-if question, not the what's-best question, and aims to predict outcomes as opposed to determining best outcomes. Simulation by itself can never say anything about the quality of the solution—it can only provide statistical analyses of possible outcomes for a fixed set of prespecified decision parameters.

Prescriptive Analysis

The role of prescriptive analytics is to provide recommendations in support of decision-making processes, where the objective is to determine a set of decisions and/or actions that gives rise to the best possible results based on various outcomes predicted by predictive analytics and subject to various constraints. This activity is also called decision making under uncertainty or optimization under uncertainty and is often based on methods from stochastic modeling and mathematical optimization.

When solving an optimization problem, it may be possible to evaluate the solution quality of each option in the decision space and select a best option. For very small problem instances, this manual enumerative approach may work well. However, for the typical optimization problems encountered in the course of P&R missions, the enumerative approach is not feasible; simply evaluating a few options within a large decision space will almost surely not find a best option and would instead render solutions that differ significantly from a best option. The general class of optimization methods addresses these difficulties by efficiently searching through the decision space of possible options in a mathematically precise manner that does not look at the vast majority of options to identify a best set of decisions or actions with respect to the desired objective and subject to any given constraints.

In addition to providing optimal recommendations, prescriptive analytics can support an elevated form of what-if and scenario analysis that explores optimal solutions (as opposed to a particular option in the decision space) across a spectrum of objectives, conditions, and inputs. For example, optimization methods might be executed to solve an initial instance of the prescriptive analytics problem at hand; these optimization methods are then executed in an iterative manner under various changes of interest in the objective, constraints, assumptions, predictive models, model parameters, and so on to ultimately provide a recommendation.

This chapter discusses classes of prescriptive analytics from the perspective of two key areas that are intimately connected. The first concerns stochastic (or probabilistic) models of uncertainty, where the goal is to model different aspects of the decision-making problem and their various sources of uncertainty, often building on top of the results of predictive analytics. The second area concerns mathematical optimization of decisions within the context of stochastic models of uncertainty of different aspects of the problem, where the goal is to either (1) provide a set of decisions or actions at the start of the time horizon that gives rise to the best possible results over the horizon or (2) provide a set of adaptive decision-making policies for dynamic adjustments to decisions or actions throughout the time horizon as uncertainties are realized. These two key areas of prescriptive analytics are discussed in the following subsections.

It is important to note that this chapter focuses on classes of mathematical prescriptive analytics methods that have been developed and applied to address problems arising in the private sector with prescriptive decision-making problems related to P&R missions. As a brief illustrative example of the application of such prescriptive analytics methods, studies addressing the problem of retention typically start with a statistical analysis (predictive analytics) of the entire workforce, looking to identify characteristics of people most likely to leave. This analysis renders models that predict losses for the different groups of people and also produces models that predict the estimated change in losses for each group as a function of the amount of additional compensation provided to that group. Then mathematical optimization under uncertainty methods (prescriptive analytics) are developed, incorporating these predictive models to determine the best policy for the amount of compensation to be allocated to reduce losses (increase retention) for each group of people composing the workforce in order to best match demand. This decision-making optimization also takes into account the various trade-offs among other workforce policy levers such as hiring (recruiting) and reskilling (training) to best match demand. One such study for a business unit within IBM led to relative revenue-cost benefits over previous approaches commensurate with 2 to 4 percent of the total revenue targets for the business (refer to Cao et al. [2011], as well as some of the examples in Appendix D, for additional details.) The classes of prescriptive analytics methods described in this chapter create opportunities to evaluate, revisit, and improve some policy decisions related to P&R missions within DoD.

Stochastic Models of Uncertainty

A prerequisite for using optimization to determine a best set of decisions or actions is the development of stochastic models of the system and the underlying processes associated with the decision-making problem. More specifically, stochastic models provide the main context for the formulation and solution of the optimization problem of a system of interest by capturing the relationships among actions and outcomes in the system, by characterizing the various sources of uncertainty in the system, and by capturing the dynamics of the system over time. In this regard, the stochastic models bring together a representation of different aspects of the decision-making problem that are sometimes taken directly from the output of predictive analytics, such as forecasts for future demand of DoD or Service personnel over time. These models are sometimes built on top of the output of predictive analytics for different aspects of the decision-making problem, such as predictions of attrition in certain personnel areas as part of a stochastic model of the availability of P&R personnel over time. These relationships, while critical, are often infused with subtleties and complex interactions.

Stochastic models characterize and predict how the system of interest and its underlying decision-making processes will behave over time under a specific set of decisions or actions. Various mathematical methods can be used to obtain such results for these stochastic models, ranging from analytical solutions to numerical computations or simulation. The most appropriate methods will often depend upon the complexity of the stochastic models and the formulation and solution of the optimization problem of interest within the context of the stochastic models. The domain knowledge needed for this area spans stochastic processes, probability theory, stochastic modeling, and simulation theory.

The interplay of stochastic models and data creates critical and complex dependencies. High-fidelity stochastic modeling of different aspects of the system of interest can be computationally intensive and require significant mathematical expertise, but it is a necessary component of the decision process. The outcome is high-quality and robust decisions that balance objectives, resource limitations, goals, processes, and organizational needs. Compromising on the quality of the underlying data, stochastic models, or use of optimization will likely lead to decisions that are less than the best—often much less.

To illustrate aspects of the points noted above, two examples of stochastic models of uncertainty involved in decision-making problems related to those of P&R are presented in Appendix D. The first example is based on the use of stochastic loss networks to model the probability of sufficient capacity or readiness of resources given uncertainty around the demand for such resources, as well as the interactions and dynamics of resources across different projects and tasks. The second example uses discrete-time stochastic processes to model the evolution of capabilities and readiness of personnel resources over time, given uncertainty around the time-varying supply-side dynamics with personnel acquiring skills, gaining experience, changing roles, and so on, including some personnel leaving and other new personnel being introduced.

Mathematical Optimization Under Uncertainty

The role of mathematical optimization is to determine a set of decisions or actions that gives rise to the best possible results within the context of the stochastic models of the system of interest and subject to various constraints. More specifically, a general formulation of a single-period decision-making optimization problem can be expressed in terms of minimizing or maximizing an objective functional of interest subject to various constraint functionals. The objective functional and constraint functionals define the criteria for evaluating the best possible results with respect to the decision variables and other dependent variables, where these and related variables are based on the stochastic models of the system of interest. The relationships among these components of the optimization formulation are critically important and often infused with subtleties and complex interactions.

In addition to optimizing a set of one-time decisions, mathematical optimization also determines a set of adaptive decision-making policies for dynamic adjustments to decisions or actions throughout the time horizon as uncertainties are realized that give rise to the best possible results within the context of the stochastic models of the system of interest and subject to various constraints. More specifically, a general formulation of a multi-period decision-making optimization problem can be expressed in terms of minimizing or maximizing over time an objective functional of interest subject to various constraint functionals. The time-dependent objective functionals and constraint functionals define the criteria for evaluating the best possible results over a given time horizon with respect to the time-dependent decision variables and other dependent variables, where these and related variables are based on the stochastic models of the system of interest. The relationships among these components of the optimization formulation are critically important and often infused with subtleties and complex interactions.

When the objective and constraint variables are deterministic (e.g., a point forecast of expected future demand, possibly over time), then the solution of the optimization problem falls within the domains of mathematical programming and deterministic dynamic programming and optimal control methods, the details of which depend on the properties of the objective functional(s) and constraint functionals (e.g., linear, convex, or nonlinear objective and constraint functionals) and the properties of the decision variables (e.g., integer or continuous decision variables). On the other hand, when the objective and constraint variables are random variables or stochastic processes (e.g., a distributional forecast of future demand or a stochastic process of system dynamics, possibly over time), then the solution of the optimization problem falls within the domains of mathematical optimization under uncertainty and stochastic dynamic programming and optimal control methods, the details of which again depend upon the properties of the objective functional(s) and constraint functionals and the properties of the decision variables. In addition, for the multiperiod case, a filtration is often included in the formulation to represent all historical information of a stochastic process up to a given time (but not future information), where both the decision variables and the system are adapted to this filtration.

Hence, mathematical optimization generally renders solutions that identify a set of decisions or actions at the start of the time horizon or identify a set of dynamic decision-making policies for dynamic adjustments to decisions or actions throughout the time horizon adapted to filtrations, in both cases having the goal of achieving the best possible results within the context of the stochastic models of the system of interest and subject to various constraints. Various mathematical methods can be used to obtain these solutions based in large part on the properties of the stochastic models of the system and its underlying decision processes. The most appropriate methods will often depend on the complexity of the underlying stochastic models of uncertainty and the details of the formulation of the optimization problem of interest within the context of the stochastic models. The domain knowledge needed for this area spans stochastic processes, probability theory, optimization theory, control theory, and simulation theory.

When point predictions are used exclusively, then mathematical programming is often most applicable, including linear programming (Vanderbei, 2013), convex optimization (Boyd and Vandenberghe, 2004), combinatorial optimization (Nemhauser and Wolsey, 1999; Schrijver, 2003; Lee, 2004), integer programming (Nemhauser and Wolsey, 1999; Conforti et al., 2014), nonlinear optimization (Ruszczynski, 2006), and deterministic dynamic programming and optimal control (Bertsekas, 2005). Otherwise, when richer probabilistic characterizations are available with analytical solutions of the stochastic models and processes, then mathematical programming, stochastic optimization (Chen and Yao, 2001; Yao et al., 2002), stochastic programming (King and Wallace, 2012), stochastic dynamic programming, and stochastic optimal control (Yong and Zhou, 1999; Bertsekas, 2012) are often applicable, with a best choice depending on the characteristics of the analytical solutions and the details of the formulation of the optimization problem. On the other hand, when simulation methods provide the only means of evaluating the stochastic models of the system and its underlying decision processes, then simulation optimization with stochastic approximation (Asmussen and Glynn, 2007; Nelson and Henderson, 2007; Dieker et al., 2016) is often most applicable, where stochastic approximation algorithms consist of a combination of gradient methods for searching the feasible solution space and simulation methods for evaluating the system at any point in the feasible space; stochastic approximation can be similarly exploited when only numerical methods are available to evaluate the stochastic system models. Lastly, when a certain measure of robustness against uncertainty of the system and its underlying decision processes can be represented as deterministic variability in the value of the parameters of the decision-making problem itself or its solution, then robust optimization (Ben-Tal et al., 2009) is often most applicable.

Another form of mathematical optimization for prescriptive analytics concerns an elevated form of what-if and scenario analysis that explores optimal solutions across a spectrum of objectives, constraints, conditions, and other aspects of the problem formulation. As an illustrative example, a mathematical optimization under uncertainty method can be used to determine the solution of an instance of the single-period optimization problem above that maximizes the objective in expectation under a given set of standard deviations for the random variables involved, and then this step is repeated for different sets of standard deviations. Each such solution provides the decisions or actions that render the best possible result in expectation of a given level of variability. In the context of modern portfolio theory, this set of solutions can be used to define the efficient frontier where each portfolio solution has the feature that there exists no other portfolio with a higher expected return and the same standard deviation of return (level of risk). More generally, this approach supports investigation of the sensitivity of the optimal solution of a mathematical optimization under uncertainty problem with respect to aspects of the formulation together with associated trade-offs, thus aiding the decision-making process and enabling basic forms of risk hedging. Laferriere and Robinson (2000) consider a related approach that exploits ideas from stochastic programming in order to compute hedged decisions; the method was implemented by the U.S. Army TRADOC Analysis Center (White Sands Missile Range, New Mexico) in a decision-support system with extensive use in Army analysis.

A formal definition and discussion of both single-period and multi-period forms of mathematical optimization under uncertainty is given in Appendix D. In addition, to illustrate aspects of the points noted above, Appendix D provides three detailed examples of mathematical optimization under uncertainty involved in decision-making problems related to those encountered by P&R. The first is based on optimization of stochastic loss networks to determine the best capacity and readiness of resources given uncertainty around the demand for such resources. The second is a stochastic dynamic program example based on optimization of discrete-time stochastic decision processes of the evolution of capabilities and readiness of personnel resources over time given uncertainty around the time-varying supply-side dynamics. The third is an example of stochastic optimal control of the dynamic allocation of capacity for different types of resources in order to best serve uncertain demand so that expected net benefit is maximized over a given time horizon based on the rewards and costs associated with the different resource types.

Optimal Solutions

It is important that prescriptive analytics methods provide the user with some basis for understanding why the optimal set of decisions or actions provided will give rise to the best possible results subject to the specified constraints. A key element for doing so involves providing the user with the predicted outcomes from predictive analytics for the optimal set of decisions or actions, as well as the predicted outcomes for alternative sets of decisions or actions for comparison.

REFERENCES

Al-Rfou, R., V. Kulkarni, B. Perozzi, and S. Skiena. 2015. Polyglot-NER: Massive multilingual named entity recognition. Pp. 586-594 in *Proceedings of the 2015 SIAM International Conference on Data Mining* (S. Venkatasubramanian and J. Ye, eds.). Society for Industrial and Applied Mathematics, Vancouver, British Columbia, Canada.

Asmussen, S., and P.W. Glynn. 2007. *Stochastic Simulation: Algorithms and Analysis*. New York: Springer-Verlag.

Ben-Tal, A., L. El Ghaoui, and A. Nemirovski. 2009. *Robust Optimization*. Princeton, N.J.: Princeton University Press.

Bertsekas, D.P. 2005. *Dynamic Programming and Optimal Control*. Vol. I. 3rd ed. Nashua, N.H.: Athena Scientific.

Bertsekas, D.P. 2012. *Dynamic Programming and Optimal Control*. Vol. II. 4th ed. Nashua, N.H.: Athena Scientific.

Biewald, L. 2015. The data science ecosystem part II: Data wrangling. *Computerworld*. April 1. http://www.computerworld.com/article/2902920/the-data-science-ecosystem-part-2-data-wrangling.html.

Boyd, S., and L. Vandenberghe. 2004. *Convex Optimization*. New York: Cambridge University Press.

Breiman, L., J.H. Friedman, R.A. Olshen, and C.J. Stone. 1984. *Classification and Regression Trees*. Monterey, Calif.: Wadsworth & Brooks/Cole Advanced Books & Software.

Burges, C.J.C. 1998. "A Tutorial on Support Vector Machines for Pattern Recognition." http://research.microsoft.com/pubs/67119/svmtutorial.pdf.

Cao, H., J. Hu, C. Jiang, T. Kumar, T.-H. Li, Y. Liu, Y. Lu, S. Mahatma, A. Mojsilovic, M. Sharma, M.S. Squillante, and Y. Yu. 2011. OnTheMark: Integrated stochastic resource planning of human capital supply chains. *Interfaces* 41(5):414-435.

Chen, H., and D.D. Yao. 2001. *Fundamentals of Queueing Networks: Performance, Asymptotics and Optimization*. New York: Springer-Verlag.

Conforti, M., G. Cornuejols, and G. Zambelli. 2014. *Integer Programming*. Switzerland: Springer International.

Cox, D.R. 1958. The regression analysis of binary sequences (with discussion). *Journal of the Royal Statistical Society, Series B (Methodological)* 20:215-242.

Dasu, T., and T. Johnson. 2003. *Exploratory data mining and data cleaning*. Wiley.

Davenport, T.H., and J.G. Harris. 2007. *Competing on Analytics: The New Science of Winning*. Boston: Harvard Business Review Press.

Dieker, A.B., S. Ghosh, and M.S. Squillante. 2016. Optimal resource capacity management for stochastic networks. *Operations Research*, submitted. http://www.columbia.edu/~ad3217/publications/capacitymanagement.pdf.

Dietrich, B.L., E.C. Plachy, and M.F. Norton. 2014. *Analytics Across the Enterprise: How IBM Realizes Business Value from Big Data and Analytics*. Indianapolis: IBM Press.

Enders, C.K. 2010. *Applied Missing Data Analysis*. New York: Guilford Press.

Fisher, K., and D. Walker. 2011. The PADS Project: An overview. In *Database Theory-ICDT 2011*. (T. Milo, ed.). Proceedings of the 14th International Conference on Database Theory, Uppsala, Sweden, March 21-23.

Friedman, J. , T. Hastie, and R. Tibshirani. 2010. Regularization paths for generalized linear models via coordinate descent. *Journal of Statistical Software* 33(1):1-22.

Getoor, L., and A. Machanavajjhala. 2012. Entity resolution: Theory, practice and open challenges. Pp. 2018-2019 in *Proceedings of the VLDB Endowment* (Z.M. Ozsoyoglu, ed.). Vol. 5, Issue 12.

Glickman, M.E., and D.A. van Dyk. 2007. Basic Bayesian Methods. In *Topics in Biostatistics (Methods in Molecular Biology)* (W.T. Ambrosius, ed.). Totowa, N.J.: Humana Press Inc.

Hastie, T., R. Tibshirani, and J. Friedman. 2008. *The Elements of Statistical Learning: Data Mining, Inference, and Prediction*. New York: Springer-Verlag.

Hill, T., and P. Lewicki. 2006. *Statistics: Methods and Applications: A Comprehensive Reference for Science, Industry, and Data Mining*. Tulsa, Okla.: StatSoft, Inc.

Isaac, D., and C. Lynes. 2003. *Automated data quality assessment in the intelligent archive*. Technical Report, Intelligent Data Understanding. NASA: Goddard Space Flight Center, Washington, D.C.

Kimball, R., and J. Caserta. 2004. *The Data Warehouse ETL Toolkit*. Indianapolis: Wiley.

King, A.J., and S.W. Wallace. 2012. *Modeling with Stochastic Programming*. New York: Springer-Verlag.

Laferriere, R.R., and S.M. Robinson. 2000. Scenario analysis in U.S. Army decision making. *Phalanx* 33(1):11-16.

Lee, J. 2004. *A First Course in Combinatorial Optimization*. New York: Cambridge University Press.

Leek, J., and R. Peng. 2015. What is the question? *Science* 347:1314-1315.

Lustig, I., B. Dietrich, C. Johnson, and C. Dziekan. 2010. The analytics journey. *INFORMS Analytics Magazine*, pp. 11-18.

NASEM (National Academies of Sciences, Engineering, and Medicine). 2016. *Statistical Challenges in Assessing and Fostering the Reproducibility of Scientific Results: Summary of a Workshop*. Washington, D.C.: The National Academies Press.

Naumann, F. 2013. Data profiling revisited. *ACM SIGMOD Record* 42(4): 40-49.

NDBC (National Data Buoy Center). 2009. *Handbook of Automated Data Quality Control Checks and Procedures*. Stennis Space Center, Mississippi.

Nelson, B.L., and S.G. Henderson, eds. 2007. *Handbooks in Operations Research and Management Science*, Chapter 19. Elsevier Science.

Nemhauser, G.L., and L.A. Wolsey. 1999. *Integer and Combinatorial Optimization*. Hoboken, N.J.: Wiley.

NRC (National Research Council). 2006. *Defense Modeling, Simulation, and Analysis: Meeting the Challenge*. Washington, D.C.: The National Academies Press.
NRC. 2013. *Frontiers in Massive Data Analysis*. Washington, D.C.: The National Academies Press.
Olguín-Olguín, D., and A. Pentland. 2010. Sensor-based organisational design and engineering. *International Journal of Organisational Design and Engineering* 1(1/2).
Rahm, E., and H.H. Do. 2000. Data cleaning: Problems and current approaches. *IEEE Data Engineering Bulletin* 23(4): 3-13.
Ruszczynski, A. 2006. *Nonlinear Optimization*. Princeton, N.J.: Princeton University Press.
Schrijver, A. 2003. *Combinatorial Optimization: Polyhedra and Efficiency*. Berlin Heidelberg: Springer-Verlag.
Smith, D., G. Timms, P. De Souza, and C. D'Este. 2012. A Bayesian framework for the automated online assessment of sensor data quality. *Sensors* 12(7):9476-9501.
Smola, A.J., and B. Schölkoph, 1998. "A Tutorial on Support Vector Regression." http://www.svms.org/regression/SmSc98.pdf.
Stein, B., and A. Morrison. 2014. The enterprise data lake: Better integration and deeper analytics. *PwC Technology Forecast: Rethinking Integration*. www.pwc.com/us/en/technology-forecast/2014/cloud-computing/assets/pdf/pwc-technology-forecast-data-lakes.pdf.
Tang, J., S. Alelyani, and H. Liu. 2014. Feature selection for classification: A review. In *Data Classification: Algorithms and Applications* (C.C. Aggarwal, ed.). Boca Raton, Fla.: CRC Press.
Tibshirani, R. 1996. Regression shrinkage and selection via the lasso. *Journal of the Royal Statistical Society, Series B (Methodological)* 58(1):267-288.
Vanderbei, R.J. 2013. *Linear Programming: Foundations and Extensions*. 4th ed. New York: Springer-Verlag.
Vapnik, V. 1995. *The Nature of Statistical Learning Theory*. New York: Springer-Verlag.
Verborgh, R., and M. De Wilde. 2013. *Using OpenRefine*. Birmingham, UK: Packt Publishing.
Wickham, H. 2014. Tidy data. *Journal of Statistical Software* 59(10).
Yao, D.D., H. Zhang, and X.Y. Zhou, eds. 2002. *Stochastic Modeling and Optimization, with Applications in Queues, Finance, and Supply Chains*. New York: Springer-Verlag.
Yong, J., and X.Y. Zhou. 1999. *Stochastic Controls: Hamiltonian Systems and HJB Equations*. New York: Springer-Verlag.
Zou, H., and T. Hastie. 2005. Regularization and variable selection via the elastic net. *Journal of the Royal Statistical Society: Series B* 67(2):301-320.

5

Privacy and Confidentiality

INTRODUCTION

When conducting analyses relating to personnel data, it is essential that privacy, confidentiality, and fairness be considered primary factors rather than, as is too often the case, being left as an afterthought secondary to an analysis's findings. This chapter provides a brief overview of relevant privacy issues. The committee's findings and recommendations are given in Chapter 7.

As a way to establish procedures for and ensure compliance with federal privacy laws, regulations, and policies, the Deputy Assistant Secretary of Defense for Health Readiness Policy and Oversight administers the Human Research Protection Program for the Office of the Under Secretary of Defense (Personnel & Readiness), referred to throughout the report as P&R.[1] Federal regulations require that ethical guidelines[2] apply when humans are used as research subjects and that the level of risk is proportionate to the potential benefit for these subjects. Furthermore, the level of review

[1] The following laws, regulations, and policies govern the Human Research Protection Program: (1) 32 CFR 219, "Protection of Human Research Subjects"; (2) 10 USC 980, "Limitation on use of humans as experimental subjects," which provides additional requirements for obtaining informed consent; (3) 48 CFR Parts 207, 235, and 252, which address requirements for the protection of human subjects involved in research conducted under contracts; and (4) DoDI 3216.02, "Protection of Human Subjects and Adherence to Ethical Standards in DoD-Supported Research."

[2] These ethical guidelines, established by the Department of Health, Education, and Welfare's Belmont Report in 1979 and codified into federal regulation in 1991, are based on three core principles: "respect for persons, beneficence, and justice."

and oversight should be commensurate with the level of risk associated with the research. The current privacy and confidentiality protections in place with government databases rest heavily on Institutional Review Board (IRB) supervision. Contracts, access control, and de-identification are some of the tools at the disposal of the board. For example, the Person-Event Data Environment (PDE) implemented by the Defense Manpower Data Center (DMDC) and the Army Analytics Group in 2006 is regulated by the Army Human Research Protection Office; data researchers are required to apply for access and explain the analyses they plan to conduct. The data are de-identified by removal of 16 of the 18 fields specified in the Health Insurance Portability and Accountability Act (HIPAA) Safe Harbor provision, and individuals are assigned randomly generated 12-character, long-lived alpha-numerical identifiers. For individual studies, secondary randomly generated identifiers are created. The mapping between an individual's long-lived random identifier and the study-specific random identifier is destroyed either immediately or as determined by the nature of the study and IRB rulings. Data are analyzed in a protected environment accessible only to authorized users, where information flow is monitored and in many cases prohibited.[3]

However, questions remain as to whether these protections are adequate, if they interfere with legitimate use, and if protections of this sort can be extended to less traditional forms of data—such as e-mail, search histories, or Facebook and Twitter postings—to enable the harnessing of these non-traditional data to address traditional P&R concerns. Questions of this type cannot be answered in a vacuum; the answers may depend on who "owns" the data of the individual in question (e.g., military personnel, their civilian spouses, their children, military retirees, reservists, the individuals with whom they communicate). The answers may also depend on the use to which the results of analyses will be put: for example, to inform policy, to be published, to identify personnel in distress. Even in these cases, questions arise about whose information may be released and to whom it may be released. For example, an e-mail monitoring system might be used to inform individuals that their writings suggest they may need professional help, which has fewer privacy implications because no third party is involved; or, instead, it might be used to alert a commander of a distressed recruit. Privacy—even for a single data analysis task—is not a one-size-fits-all problem.

LIMITATIONS OF PRIVACY-PRESERVING APPROACHES

The Fair Information Practice Principles (FIPPs), set forth in the 1973 U.S. Department of Health, Education, and Welfare report *Records, Com-*

[3] However, data may be imported and cross-referenced with government data sets, enabling linkage attacks.

puters, and the Rights of Citizens, form the bedrock of modern protection regimes, both in national and international law, such as the Privacy Act of 1974, which regulates the federal government's maintenance, collection, use, and dissemination of personal information in systems of records (Teufel, 2008; NSTIC, 2011).[4] Among these principles, for example, are the right to know what data are collected and how they are used; a right to object to some uses and to correct inaccurate information; and an obligation of the collecting organization to ensure that the data are reliable and are kept secure.

Implicit in the FIPPs, the HIPAA Safe Harbor provisions, and the de-identification procedures carried out by the PDE is a separation between PII—personally identifiable information—and other kinds of information. The U.S. General Services Administration (GSA, 2014) defines personally identifiable information as follows:

> Information about a person that contains some unique identifier, including but not limited to name or Social Security Number, from which the identity of the person can be determined. . . . The definition of PII is not anchored to any single category of information or technology. Rather, it requires a case-by-case assessment of the specific risk that an individual can be identified. In performing this assessment, it is important for an agency to recognize that non-PII can become PII whenever additional information is made publicly available—in any medium and from any source—that, when combined with other available information, could be used to identify an individual.

Such a distinction between personally identifiable information and other kinds of information is now widely considered questionable, according to the May 2014 report of the President's Council of Advisors on Science and Technology (PCAST):

> Some techniques for privacy protection that have seemed encouraging in the past are useful as supplementary ways to reduce privacy risk, but do not now seem sufficiently robust to be a dependable basis for privacy protection where big data is concerned. For a variety of reasons, PCAST judges anonymization, data deletion, and distinguishing data from metadata . . . to be in this category. . . . Anonymization is increasingly easily defeated by the very techniques that are being developed for many legitimate applications of big data.

Today, ample evidence exists of re-identification of apparently de-identified data, through methods such as multiple queries on a single data-

---

[4] As a result, FIPPs became the foundation for privacy policy principles at the Department of Homeland Security (DHS).

base or linking the data in one database with those in another, often publicly available online (Narayanan and Shmatikov, 2008).[5] Moreover, measures commonly taken to de-identify personal data, such as scrubbing or aggregating certain data fields, degrade the utility of the data (Brickell and Shmatikov, 2008; Detmar, 2012). Existing methods for aggregation of fields, such as k-anonymity, "fail to compose," meaning that they do not survive multiple instantiations, and, taken together, the k-anonymizations of multiple overlapping data sets can completely fail to protect privacy (Ganta et al., 2008).

More generally, aggregations such as contingency tables, synthetic data, data visualizations, and multiparty computations, which are apparently not about an individual, can also be problematic, as they fail to account for the fact that information derived from explicit information about specific individuals within a data set may have a profound privacy impact. Indeed, theoretical analysis yields a Fundamental Law of Information Recovery stating that "overly accurate" estimates of "too many" statistics can completely destroy privacy (Dinur and Nissim, 2003). For example, forensic analysis of protein statistics in genome-wide association studies, coupled with a DNA sample, can leak participation in a medical case group (and, consequently, the fact of having been diagnosed with a disease) (Homer et al., 2008).

Security of the data is a crucial concern, as the Office of Personnel Management knows too well in light of the June 2015 data breach that FBI Director James Comey estimated to have affected 18 million individuals (Perez and Prokupecz, 2015). Data should always be encrypted at rest and in flight. To the greatest extent possible, computations should be carried out on encrypted data.[6] The techniques of homomorphic encryption and secure multiparty computation, which permit a data analyst to run computations on data without having direct access to the raw data themselves, even when these data are shared among multiple parties, provide the effect of having the data held by a single trusted and trustworthy data curator who will carry out computations as instructed and release the results. Nonetheless, these techniques do not ensure privacy, in that they do not address the question of what can be safely released. To see the difficulties, suppose we have a perfectly secure system that gives the answers to "counting" questions such as, "How many individuals in the database are over 6 feet tall?" Let's use our system to answer questions about the House of Representatives

[5] If the publicly available data set can be downloaded, it can be brought into a data enclave, potentially enabling a data linkage attack.

[6] Much is possible even without fully homomorphic encryption, which allows arbitrary computations to be carried out but does not yet enjoy highly efficient implementation. See, for example, Dowlin et al. (2015).

and consider these two questions: "How many members of the House of Representatives have the sickle cell trait?" and "How many members of the House, other than the Speaker of the House, have the sickle cell trait?" Neither question seems invasive. But given the exact answers to both of them, the sickle cell status of the Speaker can be determined, which is a clear privacy breach. This breach happens even when everything works exactly as it should: questions are correctly answered, data remain intact, and there is no intrusion into the data set or viewing of raw data.

The example illustrates the Fundamental Law in action: the Speaker's privacy can be completely destroyed given only two perfectly accurate statistics. The fact that the House of Representatives is large is irrelevant. There is no safety in numbers, and the same attack would work if the data set consisted of all U.S. military personnel, with the Chairman of the Joint Chiefs of Staff playing the role of the Speaker of the House.

DIFFERENTIAL PRIVACY

Differential privacy, a term tailored to the statistical analysis of large data sets, together with a large body of algorithmic work for ensuring computations that satisfy this definition, addresses this last problem. Very roughly, differential privacy ensures that the outcome of any analysis is essentially equally likely, independent of the presence or absence of the data for any individual (e.g., the Speaker of the House). The likelihood is over random choices made by the differentially private algorithm (e.g., through the addition of judiciously chosen random noise). This guarantee is very strong: It says that even if the analyst knows a complete data set of n individuals and is given all the data of an n + 1st individual x, the analyst cannot determine whether the data set actually in use is D or is $D' = D \cup \{x\}$, the union of D and x.[7]

Data sets can teach us that smoking causes cancer and thereby results in a rise in insurance premiums for the smokers.[8] But this sort of impact will occur independent of whether any individual x joins or refrains from joining the data set. Differential privacy ensures these are the only forms of harm that can arise, disentangling harms that come from "facts of life" (the data set as a whole) from harms that arise as a result of participating in the data set. If the same things are learned when the database is D' as

[7] Formally, a randomized algorithm M is ε-differential privacy if for all pairs of data sets D, D' differing in one element, and for all possible events S, the probability of S, over the randomness of M, when the data set is D, is at most e^{ε} times the probability of S when the data set is D', and vice versa. Here ε is a user-specified parameter and is the measure of privacy loss. See Dwork (2011) for an introduction.

[8] Of course, the smoker is also helped: Learning that smoking causes cancer convinces the smoker to join a smoking cessation program.

are learned when the database is D, the only things x need reasonably fear are consequences of what can be learned *without his data.*

Differential privacy permits the tracking of the cumulative privacy loss under composition. Thus, it is possible to combine simple differentially private computational primitives in order to obtain privacy-preserving algorithms for complex computational tasks while minimizing cumulative privacy loss—just as in traditional algorithm design, simple computational primitives are combined in clever and creative ways to carry out complex computations while minimizing certain resources of interest, such as time, space, generalization error, and so on. Differential privacy necessarily degrades accuracy (often necessarily) because, as discussed, perfectly accurate estimates of only two statistics can destroy privacy. In comparing differentially private algorithms, a "better" algorithm is one that gives better accuracy, or uses smaller sample size, or yields better predictive capabilities, and so on, for the same degree of privacy loss. Good differentially private algorithms exist for many standard analytical tasks (Dwork and Roth, 2013).

Differential privacy is suited to statistical data analysis, especially when the data sets are large and the trends to be identified are sufficiently strong. It can provide privacy-preserving access to data when, absent this technology, access may otherwise be impermissible. For example, one may wish to permit members of the general public to request a rich class of statistics; it may be viewed as complying with privacy policies that clearly prohibit access to raw data (because the data analyst never gets such access or, unlike in the case of computing on encrypted data, even the functionality of such access), or it may be a way to permit data to be used for purposes other than those for which they were collected. However, it is not a panacea—it does not defeat the Fundamental Law of Information Recovery, and it is not the right tool for finding a needle in a haystack.

Thus, differential privacy can be used to discover, in a privacy-preserving fashion, what "typical" behavior looks like, but it is the wrong tool for finding the outlier. For example, differentially private sentiment analysis of a corpus of e-mail can safely reveal overall troop sentiment, but it cannot be used to find individuals in great distress.

In addition, the field is only now in transition from theory to practice; individual use cases[9] and academic investigations[10] are far from yielding a library of differentially private methods, and much research and algorithm engineering remain to be done. On a positive note, differential privacy protects against overfitting and false discovery attributable to adaptive data

[9] Examples of individual use cases include On the Map, RAPPOR, and Mobility.

[10] These include Penn State University's project Putting Differential Privacy to Use and Harvard University's project Tools for Sharing Research Data.

analysis, in which the second question posed to a data set depends on the answer to the first, and the 100th study undertaken on a data set depends on the outcomes of the first 99 studies of the same set (Dwork et al., 2015a and 2015b).

INSTITUTIONAL REVIEW BOARDS

IRB review has been the preferred approach within DoD to assess and control risk in permitting access to data, and the techniques discussed above are not an alternative to the IRB mechanism but represent some of the tools at its disposal. The key lesson drawn from privacy attacks, both in practice and in theory, is that information flows and combines in mysterious ways. In (pre–World Wide Web) 1980, Denning and Schlörer wrote, "These results show that existing designs for query systems do not adequately prevent disclosure of confidential data by combinatorial inference. . . . [I]t has only been recently that we have begun to understand the myriad of inference techniques that may be used. We are continually finding that compromise is easier than once thought." These problems are heavily exacerbated in the networked world—and it is precisely the web of information of this networked world that DoD wishes to constructively employ—and when the same or overlapping data sets are analyzed over and over. As a result, the task of the IRB is extremely challenging; tools and standards are needed to assist the boards, together with general principles to be applied.

The following suggestions have been made for sharing human subjects' data for research: The establishment of Safe Harbor lists specifying contexts in which specific techniques can be used without IRB review, organizational infrastructure for keeping the list current, and general principles to guide the IRB when a ruling is required (Vadhan, 2011). These suggestions are discussed briefly, noting that in addition to the Safe Harbor list, the guidance should include a "danger list" of data-sharing methods to be eschewed because they do not provide sufficient protection. Each entry in the Safe Harbor list specifies a class of data sources (e.g., electronic health records not including genomic data), a class of data-sharing methods (e.g., specific collections of aggregations or interactive mechanisms that achieve a given level of differential privacy), a class of informed consent mechanisms, and a class of potential recipients. IRB review would not be necessary in a context in which one of the Safe Harbor entries would apply.

To remain current, to evolve toward being comprehensive, to accommodate contexts that were not previously anticipated, and to take into account new developments in the scientific community's constantly evolving understanding of data privacy risks and countermeasures (which may lead to either additions or deletions from the Safe Harbor list), the Safe Harbor list should be maintained by a periodically convened task force including data privacy experts from computer science, statistics, and law, as well as members of IRBs and researchers who do various kinds of human-subjects research. This Safe Harbor list review could be done under the purview of a body such as the National Center for Health Statistics (NCHS), the National Institute of Standards and Technology (NIST), or another relevant group.

For contexts outside the Safe Harbor list, the following guiding principle should be helpful: "No individual should incur more than a minimal risk of harm from the use of his or her data in computing the values to be released, even when those values are combined with other data that may be reasonably available" (Vadhan, 2011).

In considering what is "reasonably available," the IRB should consider the extent to which the revealed values depend uniquely on an individual's data and the potential harm that may result. An IRB should ask, "Would the proposed data-sharing method be protective if the study consisted of a single individual?" A negative answer is an indication that technical expertise might be needed.

THE FEDERAL STATISTICAL SYSTEM'S LEGAL AND GOVERNANCE POLICY

The Privacy Act of 1974 and the use of IRBs may not be sufficient to control, and more importantly leverage, the vast amounts of administrative and survey data about the military population that are collected by the Department of Defense and each of the military Services. These data have huge potential for use in cross-sectional and longitudinal analyses and could provide new insights into the military population as well as have spillovers for understanding microcosms of populations in the civilian sector. For example, these data would be valuable for addressing questions about how social and organizational factors affect the behavior of individuals and their units (NRC, 2014; NRC, 2015). Unlike the civilian statistical agencies, DoD does not report the data from its data collections in a systematic way that is accessible by DoD or the public, nor does it make these data available to researchers (unless under contract to DoD). The urgency for DoD to create an infrastructure and framework for administrative and statistical data collections is highlighted by the big data revolution and increasing recognition of the importance of optimizing use of existing data sources to improve efficiency of operations and decision making (DoD, 2014). These issues have also been highlighted in this report.

For DoD to more fully use its administrative and survey data to create statistical products and to make these data available to researchers requires engaging in privacy conversations that adopt existing approaches to statistical disclosure limitations and adapt new approaches, such as differential privacy protection. There is a parallel evolution of privacy policy and governance, with a focus on access to data for research, on the civilian side of the economy. The recommendations in the National Research Council's *Private Lives and Public Policies* (NRC, 1993) played a pivotal role in shaping the U.S. federal statistical system's approach to privacy and the collection and dissemination of statistical and research data. The big data revolution has changed the privacy discussion as the availability of digital and administrative data has made massive flows of data available for research and other uses (Keller et al., 2016). DoD is an important source of these data flows to create internal and external statistical products and could benefit from joining the federal statistical system through the creation or designation of one or more DoD statistical units.

A legal framework is necessary to implement the vision to create a DoD statistical unit. The first step would be to adopt the privacy and governance structure developed by the Office of Management and Budget (OMB) in coordination with and implemented across federal statistical agencies. Two key products lay the foundation for governing privacy in the federal statistical system: (1) The OMB memorandum "Guidance for Providing and Using Administrative Data for Statistical Purposes" (Burwell, 2014) and (2) the Confidential Information Protection and Statistical Efficiency Act (CIPSEA) (Wallman and Harris-Kojetin, 2004).

The OMB memorandum is directly applicable to DoD and the PDE, and it represents an important step toward implementation of many of the recommendations set forth in this report. The memorandum has four objectives: (1) to encourage leadership to support collaboration and designation of responsibilities across programs and offices that collect statistical and administrative data and to develop data stewardship and data-quality policies; (2) to address legal and policy requirements related to privacy; (3) to discuss best practice tools, including guidance on the use of administrative data for statistical purposes; and (4) to report on progress. Box 5.1 sets forth principles based on those formulated by Burwell (2014) in her memorandum to heads of executive agencies for providing administrative data for statistical purposes, with a focus on repurposing data to minimize reporting burden and to protect privacy.

CIPSEA provides a uniform set of confidentiality protections for information collected by statistical agencies for statistical purposes, at the same time keeping in place the stringent privacy laws governing many agencies. It also provides the mechanism for DoD units to become statistical agencies or units. The law provides a standardized guide to agencies to protect the release of survey participants' information by (1) obtaining their consent at time of collection to use their data in statistical data products and (2) not exposing the information that would allow for identification of the survey respondent. The designation as a statistical agency or unit allows the desig-

BOX 5.1
Principles for Providing Administrative Data for Statistical Purposes

In her memorandum, Burwell (2014, p. 11) says as follows:

> In the context of providing administrative data for statistical purposes, both program agencies and statistical components facilitate these principles by, among other things:
>
> 1. Respecting the public's time and effort by minimizing the number of times they are asked to provide the same or similar information.
> 2. Being transparent by providing adequate notice about the planned purpose and potential statistical uses of administrative data (such as in SORNs and Privacy Act statements).
> 3. Collaborating to define which data are needed for specified statistical purposes and providing access only to those data for those purposes, and only to those who have a need for the data in the performance of their duties. Identifiable information should be provided only if the need cannot be met by relying on non-identifiable information, and even then, only relevant subsets should be provided.
> 4. Protecting data provided to the statistical agency or component against unauthorized access and disclosure. And once the data are provided to the statistical agency or component, providing the level of confidentiality protection in policy and practice necessary to ensure that the data, particularly if linked to other data, are not provided from the statistical agency or component back to the program agency for non-statistical purposes.
> 5. Implementing a set of policy and procedural safeguards, including the use of a written agreement, to certify the procedural safeguards that are employed to implement assurances of exclusively statistical uses and confidentiality. Such safeguards include applying sufficient expertise in statistical disclosure avoidance in final products in order to maintain confidentiality, taking into account risks posed by external influences such as the mosaic effect.[a]
> 6. Eliminating the identifiable information when the data are no longer needed or timely.

[a] The mosaic effect occurs when information alone is not identifiable but poses a privacy or security risk when coupled with other available information.

nee to offer the "platinum pledge" when it is collecting data with a pledge of confidentiality for exclusively statistical purposes. Under CIPSEA, only designated agencies and units can appoint "agents," who may then access the confidential data. CIPSEA is broad in that survey, interview, administrative, or other data provided to a statistical agency are protected under this statute, just as are individual data reported directly to the statistical agency (usually through surveys) (CIPSEA, 2006).

REFERENCES

Brickell, J., and V. Shmatikov. 2008. The cost of privacy: Destruction of data-mining utility in anonymized data publishing. *Proceedings of the 14th ACM SIGKDD International Conference on Knowledge Discovery and Data Mining*. http://www.cse.psu.edu/~ads22/privacy598/papers/gks08.pdf.

Burwell, S. 2014. *Memorandum for the Heads of Executive Departments and Agencies: Guidance for Providing and Using Administrative Data for Statistical Purposes*. M-14-06.

CIPSEA (Confidential Information Protection and Statistical Efficiency Act). 2006. Implementation Guidance for Title V of the E-Government Act Implementation. https://www.whitehouse.gov/sites/default/files/omb/assets/omb/inforeg/proposed_cispea_guidance.pdf.

Denning, D., and J. Schlörer. 1980. A fast procedure for finding a tracker in a statistical database. *ACM Transactions on Database Systems* 5(1):88-102.

Department of Health, Education, and Welfare. 1979. *The Belmont Report*. http://www.hhs.gov/ohrp/regulations-and-policy/belmont-report/.

Detmar, D. 2012. iDASH Workshop on Biomedical Data Sharing: Ethical, Legal, and Policy Perspectives. https://idash.ucsd.edu/events/workshops/biomedical-data-sharing-ethical-legal-and-policy-perspectives.

Dinur, I., and K. Nissim. 2003. Revealing information while preserving privacy. *Proceedings of the 22nd ACM SIGMOD-SIGACT-SIGART Symposium on Principles of Database Systems*.

DoD (Department of Defense). 2014. "Big Data: Opportunities and Challenges for Human Capital." White paper. Washington, D.C.

Dowlin, N., R. Gilad-Bachrach, K. Laine, K. Lauter, M. Naehrig, and J. Wernsing. 2015. *Manual for Using Homomorphic Encryption for Bioinformatics*. Microsoft Research Tech Report MSR-TR-2015-87. http://research.microsoft.com/pubs/258435/ManualHEv2.pdf.

Dwork, C. 2011. A firm foundation for private data analysis. *Communications of the ACM* 54(1):86-95.

Dwork, C., and A. Roth. 2013. The algorithmic foundations of differential privacy. *Theoretical Computer Science* 9(3-4):211-407.

Dwork, C., V. Feldman, M. Hardt, T. Pitassi, O. Reingold, and A. Roth. 2015a. The reusable holdout: Preserving validity in adaptive data analysis. *Science* 349(6248):636-638.

Dwork, C., V. Feldman, M. Hardt, T. Pitassi, O. Reingold, and A. Roth. 2015b. "Preserving Statistical Validity in Adaptive Data Analysis." Cornell University Library: arXiv:1411.2664. http://arxiv.org/abs/1411.2664.

Ganta, S.R., S. Kasiviswanathan, and A. Smith. 2008. Composition attacks and auxiliary information in data privacy. *Proceedings of the 14th ACM SIGKDD International Conference on Knowledge Discovery and Data Mining*. http://www.cse.psu.edu/~ads22/privacy598/papers/gks08.pdf.

GSA (General Services Administration). 2014. "GSA Rules of Behavior for Handling Personally Identifiable Information (PII)." http://www.gsa.gov/portal/mediaId/199847/fileName/CIO_P21801_GSA_Rules_of_Behavior_for_Handling_Personally_Identifiable_Information_(PII)_(Signed_on_October_29__2014).action.

Homer, N., S. Szelinger, M. Redman, D. Duggan, W. Tembe, J. Muehling, J.V. Pearson, D.A. Stephan, S.F. Nelson, and D.W. Craig. 2008. Resolving individuals contributing trace amounts of DNA to highly complex mixtures using high-density SNP genotyping microarrays. *PLoS Genetics* 4.8:e1000167.

Keller, S., S. Shipp, and A. Schroeder. 2016. Does big data change the privacy landscape? A review of the issues. *Annual Review of Statistics and Its Application* 3:161-180.

Narayanan, A., and V. Shmatikov. 2008. Robust de-anonymization of large sparse datasets. Pp. 111-125 in *Proceedings of the 2008 IEEE Symposium on Security and Privacy*. doi:10.1109/SP.2008.33.

NRC (National Research Council). 1993. *Private Lives and Public Policies: Confidentiality and Accessibility of Government Statistics*. Washington, D.C.: National Academy Press.

NRC. 2014. *The Context of Military Environments: An Agenda for Basic Research on Social and Organizational Factors Relevant to Small Units*. Washington, D.C.: The National Academies Press.

NRC. 2015. *Measuring Human Capabilities: An Agenda for Basic Research on the Assessment of Individual and Group Performance Potential for Military Accession*. Washington, D.C.: The National Academies Press.

NSTIC (National Strategy for Trusted Identities in Cyberspace). 2011. "Appendix A—Fair Information Practice Principles (FIPPs)." http://www.nist.gov/nstic/NSTIC-FIPPs.pdf.

PCAST (President's Council of Advisors on Science and Technology). 2014. *Big Data and Privacy: A Technological Perspective*. https://www.whitehouse.gov/sites/default/files/microsites/ostp/PCAST/pcast_big_data_and_privacy_-_may_2014.pdf.

Perez, E., and S. Prokupecz. 2015. "U.S. Data Hack May Be 4 Times Larger Than the Government Originally Said." *CNN Politics*. June 22. http://edition.cnn.com/2015/06/22/politics/opm-hack-18-million/index.html.

Teufel, H. 2008. "Department of Homeland Security Privacy Policy Guidance Memorandum." http://www.dhs.gov/xlibrary/assets/privacy/privacy_policyguide_2008-01.pdf.

Vadhan, S. 2011. "Comments on Advance Notice of Proposed Rulemaking: Human Subjects Research Protections: Enhancing Protections for Research Subjects and Reducing Burden, Delay, and Ambiguity for Investigators, Docket ID number HHS-OPHS-2011-0005." http://privacytools.seas.harvard.edu/files/privacytools/files/commonruleanprm.pdf?m=1365005819.

Wallman, K.K., and B.A. Harris-Kojetin. 2004. Implementing the Confidentiality Information Protection and Statistical Efficiency Act of 2002. *Chance* 17(3):21-25.

6

Commercial State of the Art in Human Resources Analytics

INTRODUCTION

Industry is leading the way in many data analytics advancements. While there are many similarities between the personnel analytics of the Office of the Under Secretary of Defense (Personnel & Readiness), referred to throughout the report as P&R, and those of industry—for example, both are intently interested in retaining specific employee skill sets and in optimizing the performance of the overarching organization—there are many characteristics unique to P&R that make it difficult to directly apply the methods used in industry. The contractual nature of servicemembers' employment and lengthy training periods are notably different from what is seen in industry. The size of the Department of Defense (DoD) workforce eclipses that of even large industry employers, therefore exaggerating the challenges faced in industry, such as developing and scaling personnel analytics software. P&R also faces legislative constraints that can make it difficult to quickly implement desired changes. Accordingly, P&R can benefit from examining the lessons learned and analytical techniques of industry-oriented analytic approaches, but it must generally adapt them to fit its unique environment.

The committee's examination of commercial-sector human resource analytics had two parts: products and practice. For products, the committee heard from a sampling of companies offering personnel analytics products (also called talent analytics or workforce analytics): IBM Kenexa, Cornerstone OnDemand, and Workday. While it is unlikely that these and similar products in their current form offer off-the-shelf solutions to DoD personnel and readiness analysis problems, there are some useful trends and insights to be garnered from this sector. On the practice side, the committee met with Google and Intel, both of which are considered to be at the forefront of applying data-intensive approaches to workforce issues, and it also researched the experiences of other companies applying data analytics to human resources (HR). While the commercial state of the art is quickly evolving, there are important lessons that may apply to DoD. The following sections describe some of the commercially available products and some of the HR analytics practices being employed in large companies.

PRODUCTS

From what the committee observed, current personnel analytics tools do not seem a good match for DoD needs, for at least a couple reasons. First, they are largely targeted at operational decision making related to HR, rather than policy-level analysis. They tend to focus on decisions around hiring, career paths, evaluation, and training and retention of individual employees, rather than on general policy questions (although Cornerstone OnDemand was deriving useful insights from the combined employment data of companies it serves). Second, the targeted market segment is not a great match. Analytics companies are typically oriented toward solving specific problems facing small- to medium-sized companies. For example, Cornerstone OnDemand focuses on initial hires of hourly workers. This segment has high turnover, which leads to a multiplicity of hiring decisions on the part of employers as well as the need to sharpen those decisions. Also, many of these jobs are easy to "instrument"—that is, to automatically collect data related to job performance, such as time cards or call-center completion rates. That market segment would perhaps contrast with active servicemembers and civil service employees in DoD, where turnover, while a concern, is much less rapid.

The development of personnel analytics products is taking place against a broader landscape of evolving enterprise resource planning tools. Those tools started out as process-centric, with different offerings for different functions (e.g., manufacturing, finance, hiring, benefits), then moved toward integrated suites of tools that span functions. Most recently, the products have become more data driven, as users try to leverage the large amounts of business data captured from internal and external sources. The commercial focus, however, seems to be on making products easier to use and easier to own, rather than on developing novel analytical methods.

On the easier-to-use front, multiple sample visualizations of analytic results to aid in choosing an appropriate visualization and automatic highlighting of interesting results are current product features, as well as "wizard" interfaces to help import data and set up analyses. Cornerstone

OnDemand emphasized the importance of integrating analytics into existing platforms (e.g., viewing attrition predictions in the same interface used to review employment applications). That point was echoed by Workday, which said it was better if the user could explore analytics results in the context of normal tools (even if the analysis itself required a larger platform) rather than having to switch to a different tool. That advice merits consideration by DoD: Look for ways the results of data analysis can be integrated into the everyday workflow of the intended consumers. The committee also heard an estimate that only 4 percent of companies were getting past descriptive and predictive analytics in the personnel domain to prescriptive analytics, and the main way to get beyond that point is to broaden the base of those who can make use of the tools or results (by, for instance, hiring more managers). That broadening, in turn, depends on easier-to-use products.

As for easier to own, vendors are making products available in the cloud rather than installing them on-site. In fact, the business model for all of Workday's products is based on cloud delivery. The advantage on the consumer side is decreased information technology costs, as there is no need for staff to install, update, and operate the product. On the producer side, it means updates are easier to roll out and can be released to all customers at once, avoiding the need to support multiple versions of a product at the same time. Presumably, the vendor also sees lower costs for support personnel that can be passed on to the customer. However, there are unique security and privacy considerations associated with cloud-based storage. This strategy may make sense for some DoD applications, where having a single server with data and analytic tools that can be accessed remotely and where security and privacy can be managed appropriately could reduce the cost of tool and data support relative to a mode of operation where analysts download data and install tools locally.

Another strategy the committee learned about from vendors is to have a common software platform that deals with functionality that all the tools from various products can use such as data storage and the user interface. Workday had a particularly flexible platform based on a triple store (Rohlοff et al., 2007) that facilitated the addition of new kinds of information.

Overall, it appears that current personnel analytics products would be more applicable at the operational functions level of the personnel offices of the various Service branches (in aid of recruiting, training, retention, or billeting) than at the policy-setting level of P&R.

PRACTICE

The committee's investigation of the literature and selected interviews reveal that a few companies are leading the way in applying data-intensive approaches to personnel operations and policy. Many companies have built up specialized groups for personnel analytics and are trying to base decisions on the actual conditions in their own companies rather than on boilerplate "best practices" or prior research (Derose, 2013). The range of conditions considered is quite broad; they include structuring of the interview process, identifying the attributes of successful (and unsuccessful) managers, setting family leave policy, challenging common knowledge within the organization, determining the key factors that affect retention, studying how patterns of communication affect unit performance, and charting effective career paths. The increasing ease with which personnel data can be archived across time has allowed more longitudinal studies to identify the most reliable indicators of employee behavior. For example, Intel found that the answers to eight survey questions plus the time accrued at the company sufficed to accurately predict retention.

One of the highest profile studies in the personnel analytics domain was Project Oxygen at Google (Bryant, 2011). There had been significant doubt in Google about the value of engineering managers, and at one point the company had gone so far as to eliminate project managers because it had no proof that they had had any impact. Project Oxygen started by assembling a large base of observations on managers from existing sources, such as performance reviews and feedback surveys. Those were supplemented by hundreds of hours of interviews with managers (which required hand coding). Analysis of the data revealed eight behaviors of good managers, such as expressing interest in employee success and well-being and helping with career development. It also revealed pitfalls, such as spending too little time on managing and communicating. (Interestingly, technical expertise ranked last on the list of eight.) The results were used to revise employee-training practices and institute coaching for low-performing managers.

Some of the trends the committee observed and advice it received on personnel analytics are highlighted below.

Mythbusting

The committee learned of multiple instances where data analytics were used to confirm (or not) the common knowledge in an organization or to challenge the status quo (the latter was colorfully referred to as "poking the bear"). Two specific examples the committee heard of myths that had been refuted by analysis of actual data were that (1) employees at headquarters are promoted faster and (2) job candidates with multiple employments in the last 5 years or who have been unemployed for a prolonged period of time are less likely to stay with a new position. In reality, employee attrition and retention are much more heavily influenced by colleagues (think of the damage that can be done by a "toxic" coworker).

In other cases, data analytics suggested alternative personnel strategies. For example, observing unusual but successful job moves suggested alternative career pathways for some employees. Another study revealed that a small segment of the workforce was happy to work on a series of short-term projects, leading to a pilot program for "internal freelancers," who are assigned to projects for a few months at a time and who can tolerate breaks in employment.

Combination of Skills

The committee saw that employees in Personnel Analytics groups (sometimes called People Analytics) often possess a variety of skills and backgrounds. In addition to data scientists, there are experts (often PhDs) in HR domains such as psychology and organizational theory as well as staff (generally MBAs) focused on different HR functions: hiring, benefits, and training. A particular study or analysis might originate with an MBA interacting with a functional group to figure out a current issue or question. The appropriate domain-area expert might design a study to answer the question, deciding which existing data would be applicable and what new information would need to be collected. A data scientist might extract and prepare the data and run the appropriate analyses, which would be interpreted by the domain expert and then be communicated back, perhaps through the MBA, to the HR client group.

Communication

The committee observed the importance of putting analysis results in a form that an end user can understand. It was stressed that correlations and trends were not in and of themselves generally actionable by decision makers. The results needed to be explicable before decision makers felt comfortable recommending changes in procedures or policies based on them. Especially when machine learning is involved, where the computation techniques can be somewhat opaque, there is a need to investigate further to develop cause-and-effect explanations. Also, it helped to invest a group leader in the results by briefing them on the results and having them present the results to the rest of the group. The committee also learned that in order to maintain high response rates to information-gathering mechanisms, it was important for employees who had answered surveys and responded to such internal communications to know that their input was being noted and acted upon.

Launching a Personnel Analytics Effort

Building a successful personnel analytics capability is not trivial. In addition to finding appropriate staff in the face of stiff competition for data scientists in particular, developing the appropriate data collection enterprise takes time and organizational will. The committee observed the need for substantial foundational work within the organization on what data mean. In one case, for example, simply coming up with a uniform definition of "full-time equivalent" took over a year of discussion among business units (engineering, sales, finance, etc.). Some of the committee's interviewees also stressed that the first projects needed to be creditable when starting a data analytics effort—initial efforts needed to show success and value.

Appropriate Techniques

Not every study requires huge amounts of data and sophisticated methods. For example, a simple spreadsheet-based model was sufficient to do a multiyear forecast of workforce shape (number of employees at each level) from current employee counts and rates of promotion, hiring, and retention. It revealed the company would shortly become top-heavy with managers if current practice continued, and it led to changes in promotion rates and policies on external hires at upper levels.

Nontraditional Data Sources

All the analytics organizations the committee studied were looking at extending their data assets beyond what they could obtain from internal business-data processing systems. For example, text analytics capabilities (discussed in more detail in Chapter 4) are reaching a level of maturity such that they are suitable for preparation and analysis of unstructured text data.

Natural Experiments

Many in industry are noting the value of natural experiments in the analytics domain. For example, studying the performance of different teams over time revealed dimensions that were highly predictive of team success, such as how long a team has been together and who is leading it.

REFERENCES

Bryant, A. 2011. Google's quest to build a better boss. *The New York Times*. March 12. http://www.nytimes.com/2011/03/13/business/13hire.html.

Derose, C. 2013. How Google uses data to build a better worker. *The Atlantic*. October 17. http://www.theatlantic.com/business/archive/2013/10/how-google-uses-data-to-build-a-better-worker/280347/.

Rohloff, K., M. Dean, I. Emmons, D. Ryder, and J. Sumner. 2007. An evaluation of triple-store technologies for large data stores. Pp. 1105-1114 in *On the Move to Meaningful Internet Systems 2007*. OTM 2007 Workshops. Berlin Heidelberg: Springer-Verlag.

7

Identifying P&R Opportunities and Implementing Solutions

INTRODUCTION

This study has covered the current state of personnel and readiness data and uses and has opened a window onto the application of analytics in the personnel and readiness environment. In the preceding chapters the committee described the challenges it observed and discussed the data and analytics methods used in industry and by other players. The study's charge was to develop a roadmap and implementation plan for integrating data analytics in support of decisions within the purview of the Office of the Under Secretary of Defense (Personnel & Readiness), referred to throughout the report as P&R. The committee interpreted that charge as calling for high-level conceptual advice, because P&R must have the latitude to craft specific plans for strengthening its capabilities in, and use of, data science.

Accordingly, the committee believes that the first step is for P&R to develop a data and analytics framework, in accordance with the advice in this chapter, to better address the short- and long-term needs of the Secretary of Defense. Many of the Department of Defense's (DoD's) data sets have been built up in response to disjointed and one-time questions (e.g., from Congress) or to deal with narrow transactional issues. They are not typically the result of a data and analytics framework derived from a coherent picture of what the Secretary of Defense and others need to know to manage DoD.[1] This chapter articulates some short-, medium-, and long-

[1] The Defense Readiness Reporting System (DRRS) is somewhat of an exception, although even it is mostly the product of federating existing data systems, both to keep costs in check and to encourage participation and compliance.

term goals to help P&R develop this framework: (1) improve data quality and sharing, (2) enhance data science methods, and (3) strengthen data science education. The first goal, data quality and sharing, can be improved immediately, while data science methods can be enhanced in the medium term. Strengthening workforce capabilities in data science can begin now but is a longer-term task.

IMPROVE DATA QUALITY AND SHARING

Data Quality

DoD has made significant investments in improving its use of data and analytics for recruiting and pay over the past 40 years,[2] by means such as sanctioning the creation of the Defense Manpower Data Center (DMDC) to provide a common set of data records and turning to the federally funded research and development centers (FFRDCs) for analytic assistance. These investments have helped DoD overcome difficulties, and arguably they have played a significant role in its long-term success. To continue that success, the analyses and the analysts supporting the Under Secretary of Defense for Personnel & Readiness need insight into the outlooks and motivations of personnel. However, these attributes may not be revealed by administrative records, and new data may be required. These new data needs must be defined and characterized, and the design of enhanced data systems must take into account how the data will be used, including their analytic applications.

One important consideration in planning enhancements to the data systems available to P&R is that inaccuracies and incompleteness often impede the full use of data in support of P&R decision making. For example, analysis of the burdens of deployment requires DMDC to work with a number of variables to determine whether an individual was serving in forward combat areas, but those variables do not always agree. Likewise, using the central records to review language competency will reveal that

[2] The impetus for many of these investments was the return of the United States to its tradition of relying entirely on paid volunteers to staff its military establishment. Recognizing that the shift from conscription to a reliance on volunteers required key decisions about compensation and the active management of recruiting, the nation's leadership turned to outside analytic expertise in support of the Gates' Commission, which buttressed President Nixon's decision to end the draft. A symposium hosted by the Center for Naval Analyses in 2014 honored Walter Oi and his influence in establishing the all-volunteer force. The website for that symposium is https://www.cna.org/news/events/all-volunteer-force, accessed January 6, 2016. Emblematic of the community of interest that these decisions created is the series of commemorative volumes produced at each of the 10-year anniversaries of the all-volunteer force in its first three decades (Bowman et al., 1986; Fredland et al., 1996; Bicksler et al., 2004).

such information is often missing. The same will be true for indicators that officers have completed joint duty assignments (that is, assignments involving more than their parent service, which ultimately can be important for promotion), because such data are often kept in separate systems. Even for data fields that are well populated in the original submissions, DMDC reports that it expends considerable effort to clean the data to ensure their accuracy.

Although many of the data sources P&R relies on arise from administrative needs, few rest on transactional record systems that may have the most relevant information. Readiness data, for example, do not typically derive from parts ordering or repair transactions but derive rather from reports on equipment availability. Biases in reporting may occur when the underlying variable is judgmental in nature; for example, whether a piece of equipment is mission capable can reflect the judgment of a commander. Likewise, deployment data generally have not depended on records of arrivals in and departures from theater but rather on a downstream transaction (the pay record). Readiness data have long suffered from this weakness, which in part led to the establishment of DRRS, with its emphasis on access to the underlying data files. Other data sources (responses to questions on race and ethnicity, say) can potentially be affected by bias, especially data collected via surveys. Transactional data may be used at the Service level but are much less likely to be used by P&R. An important exception to this generalization is health care data, which this study did not examine in detail.

Most of the data sources easily available to P&R are snapshots of either current or retrospective conditions. For example, P&R does not have all Service forecasts of enlistment contracts with which to gauge the status of recruiting, although it is conceivable these could be generated either from separate systems maintained by the individual Services or from questions asked during military entrance processing station (MEPS) testing. Rather, P&R receives data on actual enlistment results for the month as well as the status of enlistment "reservations." The latter are tallies of those who have signed contracts to enlist at a future date, the so-called Delayed Entry Program (DEP); actual enlistment results and the size of the DEP are used as recruiting barometers. Some surveys offer direct forecast information (e.g., reenlistment intentions) or indirect information (e.g., the Youth Tracking Survey, with its questions on how American youth feel about military service), but these fall short of what P&R really needs.

Individuals might enrich a number of the data sources or, at a minimum, improve their accuracy and completeness, provided there are sufficient incentives for them to offer corrections. Systems that encourage individuals to self-report accurately constitute a potentially rich source of data for P&R on both program preferences and the outlook of servicemembers on their military careers and have strong predictive power. The Army's pilot study "Green Pages," which ran from 2010 to 2012, provided insights into the preferences of mid-career officers for assignments as well as additional detail on their backgrounds and accomplishments not otherwise available. (In the pilot study they provided resumes with much more detail than is contained in their administrative personnel records.) Likewise, the Navy administers Assignment Incentive Pay, in which a modest number of hard-to-fill enlisted assignments are put up for "bid," and individuals may specify the additional compensation, if any, they would require in order to accept a particular assignment. This process implicitly provides useful information about enlisted preferences that could conceivably be used for predictive purposes such as examining trends (Carter, 2015). Self-reporting has long been relied upon to generate longitudinal data series, such as NLS79, the Millennium Cohort Survey, and the new Military Family Life Survey. These longitudinal series are often the most valuable sources of information on behaviors.

Although the administrative process, self-reported or otherwise, might generate some of the underlying data promptly, it can take considerable time before those data appear in files that the policy analyst can employ. For example, the time DMDC invests in assembling and cleaning data from the central personnel records results in a lag in the analysis compared to when the events actually took place. Likewise, although surveys are now generally administered via the Internet, the process of collecting, cleaning, and organizing the data usually takes weeks if not months, again introducing latency in the analytic process. And if the data are to be employed by an outside organization, there may be additional delays in accessing them.

DoD has an opportunity to get more value out of existing data collections by creating an expectation of data repurposing and analytics reuse. P&R could capitalize on some of the transactional data available to the Services by establishing partnerships with them. For example, repurposing assignment orders could provide a portrait of one element of future readiness (e.g., how well trained and experienced are those who are moving to a given unit?), as well as be an indicator of unit turnover. Data from tests administered at the conclusion of training could be repurposed to monitor training effectiveness and predict future unit readiness.

The assessment of noncognitive attributes through, for example, the Army's Tailored Adaptive Personality Assessment System (TAPAS) (Stark et al., 2014) offers research and operational opportunities to improve selection and placement decisions that have historically been grounded in the administration of and reliance on cognitive ability tests. Data collected could be used at the point of entry into military service as well as at mid-career to guide further investment in human capital. New monitoring elements could also be devised (end-of-tour questionnaires that would expand on the health-related questionnaires now required of those returning from a combat theater).

Converting unstructured or semistructured data from a variety of sources into structured form could be another avenue of potential value for P&R analytics. For example, natural language processing could provide valuable and timely information from structured records (documents, messages, reports). Annotating existing data with additional information—for example, geocoding—could allow new classes of tools to be developed and used.

The ideas suggested above are examples of new or repurposed data sources that P&R might explore and incorporate into its data and analytics framework. In each case, P&R will need to assess in detail whether the potential value mentioned is feasible and an important priority. It will also need to evaluate whether there are technical or management hurdles to overcome. These details, which can be quite nuanced, were not readily available to the committee and would best be examined by experts within P&R.

Finding: Despite the substantial amount of data available on DoD personnel, the data may not be appropriate for DoD's analytic tasks, or they may necessitate considerable investment in constructing the variables of interest.

Finding: Analyses developed to support the Secretary of Defense are often disjointed, one-off activities undertaken to respond to immediate questions and may lack a plan for future use of data or analytic methods.

Finding: The reuse of operational data for analytic purposes can expose issues in data collection, recording, transmission, cleaning, coding, and loading. Problems are often not detected until the point of analysis, when anomalies crop up in results.

Recommendation 1: The Office of the Under Secretary of Defense (Personnel & Readiness) should develop a data and analytics framework, and a strategy to implement that framework, that addresses both the principal outcomes of its responsibilities and the short-term and long-term needs of the Secretary, based on the findings, recommendations, and discussions outlined in this report and in the Force of the Future proposals.

Data Sharing

While P&R enjoys considerable access to data that help it address the issues for which it is responsible, substantial gaps remain, particularly for events that occur outside of DoD. Publicly available data (e.g., from social media) and data maintained by other government agencies (e.g., Department of Veterans Affairs [VA]; the Employment Cost Index and unemployment data from the Bureau of Labor Statistics; the American Community Survey and the Longitudinal Employer–Household Dynamics from the Census Bureau) could all play pivotal roles in filling current information gaps. Many Services collect information on their members that is not reported to P&R (from, say, new-recruit surveys). The advent of the capacity to analyze large data sets has made it possible to utilize such previously unused data to inform other predictions.

One example of the potential benefit of sharing data arises in the Career Intermission Program that Secretary Carter has endorsed, which allows some servicemembers to take a 1- to 3-year sabbatical (Losey, 2016; Schechter, 2016; Serbu, 2016). Despite its appeal in principle, in practice the Services have struggled to realize the program's potential. Understanding the problems each Service has encountered in implementation, and especially the experience of the Navy (the original proponent of the authority and the first to employ it), would avoid the missteps and identify early on what must be done if the program is to succeed. While longitudinal data are often essential to understanding both behavior and policy choices, few of P&R's data sources are constructed originally in that form. Rather, the analyst or the data agency (e.g., DMDC) will construct the longitudinal data set from the available "snapshot" files. The two big exceptions to this generalization are the Millennium Cohort survey (conceived as a longitudinal effort from the beginning, to track the health experiences of military personnel after the illness controversies that followed the first Persian Gulf War) and the earlier National Longitudinal Survey of Youth 1979, in which DoD participated.

A particular opportunity arises in the data collection efforts of the VA. VA data illuminate other aspects of an individual's experience after leaving active duty (pursuit of further education via the GI Bill, treatment of medical issues that may derive from service, etc.). Harmonizing the data elements from DoD and the VA would benefit both agencies, by giving DoD a better understanding of the status of reserve component members and, for the VA, by probing the antecedents of postservice health issues. More important, it would benefit veterans and provide them with the recuperative support they have earned.

Understanding the life trajectory of military personnel after they leave active duty—because they are discharged, retired, or demobilized (i.e., Reserve Component members, including those who continue to serve in a Reserve capacity)—could give P&R valuable information on the appeal of the military service or its detractions (Wilmoth and London, 2013). Unemployment data are collected by the Department of Labor without any regard for DoD's post-Service experience needs. Understanding the life trajectory would require larger sample sizes (to shrink the error associated with the estimates), different age range aggregations, knowing the reason for leaving active duty (e.g., completion of active duty obligatory service or discharge before completing obligatory service), and a marker for whether the person is enrolled in school.[3] DoD needs an interagency mechanism to persuade its sister agencies to undertake necessary changes, including a vehicle for funding when additional resources are required.

The transactional and survey data sources compiled by DoD are a valuable resource and should be maximally leveraged to support P&R policy decision making. There are legal, regulatory, cultural, and technical barriers to (appropriately and responsibly) sharing data. One particular challenge is balancing the ability to link disparate data sources for individuals against the simultaneous goal of protecting their privacy. On the technical side, choices of hardware and software architectures provide different mixes of advantages and challenges with respect to data access and management, and these trade-offs need to be considered in developing a data and analytics framework. In addition, because some of its most important data come from series established by other agencies, it behooves P&R to take a more active role in interagency data deliberations and to advocate for sample sizes and data constructs that better serve its needs. DoD will need to be willing to finance the additional costs of these requests.

The use of social media and other outside data in support of P&R decision making is worth further examination. Although DoD warns its personnel to avoid posting information regarding their responsibilities online, many individuals maintain accounts with some degree of public access. Government monitoring of such postings, while potentially of significant value, raises difficult questions about privacy and the appropriate role of government itself. For personnel who hold security clearances, the government already asserts some authority over private matters (e.g., drug tests, reporting of law enforcement actions). Internet postings are by definition public, but some may find their collection and analysis by the government unsettling. Further, there is a difference between analysis of posts by an outsider and the analysis of posts by DoD, which has private information about those making the posts. The latter situation ranks much higher on the "creepiness" scale (Tene and Polonetsky, 2014). Privacy-preserving data analysis tools, such as differential privacy, may be particularly helpful should P&R pursue analyses that draw from social media postings.

Whether it exploits social media or government records, P&R should explore whether contemporary text analysis techniques would allow it to

[3] In many states, veterans are permitted to collect unemployment insurance at the same time they are pursuing schooling under the GI Bill. A number of National Guard members are students who presumably are returning to school.

understand the content in documents of interest. Could text analytics that allow the analysis of Internet postings help P&R better understand the burdens of deployments? Could such techniques, in a further example, give P&R the ability to evaluate, fairly and even-handedly, the joint duty experiences of those applying for credit? Not only might such text analytic methods provide an efficient approach to current document reviews, they might also make available new sources of data contained in documentary sources too voluminous to examine with other methods.

Improving data sharing across Services could also prove valuable for P&R. However, standards specifying how data are collected are not comprehensive and systematic across Services. Each Service may collect similar data but utilize them differently based on Service-specific needs. This divergence results in Services tracking different data and metrics and making different policy decisions. To deal with multiple sources of data originating from multiple organizations, the strategy is to develop a process for agreeing on what new information to record, to set standards for how modifications should be made, and to render information into a common representational form.[4] As noted in Chapter 3, however, the current situation represents a clear improvement over the situation that existed before DMDC's creation.

DoD's internal processes typically do not involve data sharing but rather the reporting of results based on data. It is for that reason that in formulating a data and analytics framework as recommended in the previous section, a "preparation instruction" should be issued that tells subordinate personnel what data they must submit. Historically, that instruction was a matter of considerable debate. Perhaps the biggest implementation challenge is the difference between the philosophical outlook of the "big data" community (in which raw data are generally available for authorized analysis) and the "command and control" ethos of DoD, in which information is indeed power. There is a notion that access by more senior personnel of the bureaucracy to the underlying data sets might prove contentious, to say nothing of access by any "outsiders." The Services are reinforced in their preference for limiting access to data by their legal charge from Congress to organize, train, and equip military forces, while the role of the Secretary of Defense and his staff is to set policy and be politically responsible for the results.[5] The desire to control data needs to be balanced against policy needs.

Improving data sharing with FFRDCs could accelerate their research processes and allow enhanced opportunities for verification and validation

[4] This process usually consists of establishing a cross-Service working group, often with representation from P&R, to reach agreement.

[5] Congress has subsequently said that the grant of authority to a subordinate element of the department in no way precludes the Secretary from exercising it.

of study findings. Some of these data access challenges are discussed in the following section on Institutional Review Boards, and P&R may benefit from exploring opportunities to ease data transfer.

While those outside the federal government, especially the FFRDC staffs that are working for DoD, may request access to data available to P&R, there is no tradition of public use of data sets as, for example, characterizes the Census Bureau. DoD is starting to explore this possibility. The Person-Event Data Environment (PDE) that the Army Analytic Group and DMDC created is the principal current focus, although there are major challenges in making the PDE a widely useful tool. However, technical and cultural challenges (such as possible data reidentification and other privacy compromises), a slow and complicated approval process to gain access, lengthy reviews for data import and export, limited computational capabilities, concerns about data quality and comprehensiveness, and concerns about data ownership rules pose a significant deterrent to utilizing the PDE. In addition, it is not clear that the architecture scales up in such a way that it can serve all of P&R's needs, and forcing analysts to work through the PDE personnel (who then must work through the data owners) may represent a barrier between the analyst and the raw data. The substantial efforts undertaken by PDE personnel to prepare the data for linkage are not transparent and may inadvertently impact the results of analysis results. The Defense Health Agency has also begun exploring the creation of a data set to which greater access might be granted.

From the policy analyst's perspective, perhaps the most important innovation would be a forum or mechanism that would channel feedback from the user about the data constructs that are needed for typical analytic tasks and the nature of the variables that would best meet those needs. Some entity, such as an Office of People Analytics or an expanded role of DMDC, with clear lines of responsibility under the Confidential Information Protection and Statistical Efficiency Act (CIPSEA), could provide such a forum. The PDE could become a central activity of such an office. Such a focus could immediately help P&R leverage better data and data analytics to support what the military cares about. Services alone are not able to keep pace with the growing need for expertise in data science. Secretary Carter's Force of the Future effort calls for such an office. It could connect to similar activities across the federal government, leveraging the best analytics, capabilities, and talent.

> **Finding:** The existence of DMDC and a unified personnel file has greatly enhanced the ability of the Office of the Secretary of Defense (OSD) to understand the behavior of its personnel and to refine its policies so as to improve both retention and performance. The creation of the Civilian Personnel Data System was a similar achievement.

Finding: There are benefits to be gained by enabling deeper and richer collection and sharing of data, which support a richer picture of the individual. This could in turn allow for better matching of personnel to the needs at hand (e.g., with regard to desired data skills, language proficiencies, and experiences), improved identification of at-risk servicemembers, enhanced management of the force in terms of retention and training, and many other benefits.

Finding: The challenges of data sharing and repurposing are significant; in particular, different data definitions and formatting complicate data merging and linking. Business practices (e.g., methods, procedures, processes, and rules) vary from Service to Service and from one database to another.

Finding: Enhanced data sharing within DoD, across agencies, and with the research community at large could promote the creation of new statistical methods, tools, and products.

Finding: The existence of alternative data sources, such as social media, especially when they are tied to extensive information about individuals, may deliver deep insights relevant to the mission of P&R. Owing to concerns about privacy and appropriateness and to the difficulty of ensuring statistical validity, further pursuit of this path requires careful consideration and additional research.

Recommendation 2: The Office of the Under Secretary of Defense (Personnel & Readiness) should investigate the feasibility of exploiting alternative data sources to augment traditional methods for measuring collective sentiment, evaluating recruitment practices, and classifying individuals (for creditworthiness, perhaps, or for battle-readiness). Hand in hand with this effort there should be an investigation into privacy technology appropriate for these scenarios for data use.

Recommendation 3: The Office of the Under Secretary of Defense (Personnel & Readiness) should identify incentives to enhance data sharing and collection, such as the following:

- Tracking usage of data by source in repositories such as the Person-Event Data Environment and periodically reporting back to data providers on usage (e.g., number of uses, who the users are, the nature of the study, or analysis the data contributed to);

- Providing incremental funding on contracts that involve data collection and organization to cover the costs of archiving and documenting the data for other users; and
- Giving preference to projects for constructing or redesigning operational data systems that include explicit functionality to support data sharing.

Recommendation 4: The Office of the Under Secretary of Defense (Personnel & Readiness) should leverage opportunities to improve access, including better reuse of prior data, tools, and results, and should investigate incentives to increase interagency and inter-Service data sharing.

Recommendation 5: The Office of the Under Secretary of Defense (Personnel & Readiness) should establish a working group with representation from the Services and other elements of the Department of Defense, as appropriate, to

- Identify productive new fields and formats for personnel files, such as enabling the inclusion of unstructured data and free-form text in future records;
- Identify opportunities for data sharing between Services and the Office of the Under Secretary of Defense (Personnel & Readiness) and within Services and lower barriers to such sharing;
- Work with organizations that provide operational data or collect them for analysis to improve data quality by providing standard ways for data users to report problems with data collections and channel those reports back to data providers when appropriate;
- Clarify self-reporting rules and practices;
- Identify legal and regulatory barriers to the appropriate and responsible sharing of data; and
- Examine new hardware and software architectures that facilitate data access and data management.

Finding: The development of the Person-Event Data Environment is a positive step in making some data more easily accessible. However, certain technical and cultural factors deter the use of this tool.

Benefits

- Spreads the overall cost of data acquisition, cleaning, ingestion, and linking.
- Reduces time for researchers identifying and downloading data, since they work on it in situ.
- Aims to improve handling of sensitive data.

- Monitors data usage.
- Creates a group that supports users with data and tool issues.

Drawbacks

- Sensitive personally identifiable information is susceptible to reidentification and other privacy compromises such as revelation of sensitive traits or attributes.
- Linkage attacks—innocuous data in one data set used to identify a record in a different data set containing both innocuous and sensitive data—can be carried out via external data sets brought into the PDE by researchers.
- Review processes are lengthy for access to some data.
- Delays in the review process for export of analysis results pose a deterrent to publication and peer review.
- The hurdles to become a PDE user mean that the current user community is much smaller than intended.
- Some users have been limited by the computational power, memory, and tools of the current installation.
- The PDE does not solve completeness and quality issues in the underlying data sources.
- There does not exist a systematic mechanism for reporting data problems.
- Some PDE users say they have been given conflicting statements about the ownership of external data uploaded into the PDE.

Recommendation 6: The Defense Manpower Data Center should assess how well the Person-Event Data Environment is working and whether it is serving its intended community. In doing so, the center should consider taking the following steps to improve the usability of the Person-Event Data Environment and enhance its value:

- Assess if current privacy and security policies are adequate, taking into account modern methods of attack and sources of auxiliary information that can aid in these attacks, such as multiple releases of statistics and data sets (Ganta et al., 2008), linkage attacks that make use of public sources (Sweeney, 1997; Narayanan and Shmatikov, 2008), and chronological correlations with public sources (Calandrino et al., 2011).
- Analyze data usage information, both for privacy and determining value of assets.
- Do a better job of establishing and defining a user community for knowledge sharing. This includes improving relationships with the federally funded research and development centers doing work

for the Department of Defense and determining which researchers would benefit from the capabilities of the Person-Event Data Environment.

- Remove unnecessary barriers for researchers to gain access to the system.
- Enhance computational power, memory, and tools.
- Respond to concerns about the quality and comprehensiveness of available data.
- Develop an explicit process for reporting data problems.
- Clarify data ownership rights to external data that are uploaded and merged.
- Assess protocols for accessing personally identifiable information.
- Review approval process for exporting analysis results.
- Consider widening access to the data and/or rebalancing Institutional Review Board requirements by establishing a differentially private interface.[6]

Institutional Review Boards

Institutional review boards (IRBs) are normally involved in carrying out human-subjects research in order to ensure that the work is handled ethically. The human-subjects protection regulations promulgated in 1981 (45 C.F.R. § 46) and the revisions known as the "Common Rule," issued in 1991, aim at delineating human-subjects research policies, and they apply to FFRDCs and other groups that work with DoD on P&R research projects; these regulations do not apply to P&R data analyses completed solely for operational purposes. For most studies that assemble enough human data to infer useful results, IRB oversight is necessary.

The committee was told by a number of analysts from those nonfederal research organizations that they are sometimes required to satisfy often multiple reviews by IRBs in both their own institutions and DoD, and they often feel this duplication is unnecessary. In addition, there is sometimes a lack of clarity as to which DoD IRB should conduct a review of protocols. In at least one case, a local commander successfully insisted that a study gathering data from his command had to be approved by his IRB, even though IRBs from both DoD and the FFRDC had already approved the study. The committee was told how this duplication adds steps and considerable time to an already lengthy process. Sometimes an IRB may require changes that then need to be resubmitted to and reviewed by other IRBs, further extending the timeline.

Similar problems have been noted by others, and the 2014 National

[6] For a discussion of differential privacy, see Chapter 5.

Research Council (NRC) report *Proposed Revisions to the Common Rule for the Protection of Human Subjects in the Behavioral and Social Sciences* includes numerous recommendations for improving IRB processes. For example, it recommends establishing single IRBs of record for multisite studies.

Another kind of barrier is the requirement imposed by the Paperwork Reduction Act that U.S. government agencies obtain approval from the Office of Management and Budget (OMB) before collecting information from groups of more than nine persons. The committee was informed that in some cases FFRDC analyses using focus groups had been conducted with information gathered from only nine people, when a desirable sample size would have been considerably larger, because it was thought to be impractical to obtain OMB approval within the time allotted for the study.

Finding: Reviews by multiple Institutional Review Boards can significantly slow down the research process and add months or years to the time it takes for researchers to have access to DoD data. This creates a serious problem for responding to policy needs in a timely manner.

Recommendation 7: In order to support timely and efficient research, the Office of the Under Secretary of Defense (Personnel & Readiness) should encourage streamlining of Institutional Review Board processes that involve multiple organizations—for example, federally funded research and development centers and the Department of Defense.

Recommendation 8: The Department of Defense should carry out research on the feasibility of differential privacy methods for its personnel analytics. These methods could reduce the need for Institutional Review Board oversight.

Privacy and Confidentiality

As discussed in Chapter 5, the Fair Information Practice Principles (FIPPs) that regulate the federal government's maintenance, collection, use, and dissemination of personal information in systems of record outline that an individual has a right to know which data are collected and how they are used, as well as to object to some uses and to correct inaccurate information. The collecting organization for its part must ensure that the data are reliable and kept secure. As data collection on individuals continues, and as data sharing becomes more common, these are important principles for DoD and P&R to keep in mind.

One way to do this is for DoD to adopt or adapt the privacy and governance structure developed by the Office of Management and Budget for civilian statistical agencies. The Confidential Information Protection and Statistical Efficiency Act (CIPSEA) provides a uniform set of confidentiality protections for information collected while adhering to stringent privacy laws governing many agencies. It serves as a standardized guide for agencies to protect the release of survey participants' information, both by obtaining their consent at time of collection to use their data in statistical data products and by protecting the information that would allow for identification of the survey respondent.

> **Recommendation 9:** The Department of Defense should consider adopting or adapting the privacy and governance structure developed by the Office of Management and Budget for civilian statistical agencies. In particular, the department should follow the guidance on use of administrative records and establishing of statistical units under the Confidential Information Protection and Statistical Efficiency Act for both military and civil service personnel. In doing so, the department should examine the applicability of Fair Information Practice Principles in the treatment of Defense Manpower Data Center data.

> **Recommendation 10:** The Defense Manpower Data Center, in its role as steward of the Person-Event Data Environment, should consider ways to adapt and use privacy and governance practices that the Office of Management and Budget has created for civilian use.

ENHANCE DATA SCIENCE METHODS

Currently, barriers exist to the effective use of analytic methods that support P&R's policy decision making. For a variety of data sources and systems, stronger analytic approaches are needed to shorten the time required to go from raw observations to analyses to decisions. The committee found an uneven use of data science throughout the DoD personnel offices, with some areas having advanced skills and others just beginning to incorporate entry-level analytics to inform decision makers.

The potential for strengthening and expanding the use of data science methods in P&R aligns well with Secretary Carter's recent comments on the Force of the Future. There, he outlined a number of goals that have a direct relationship to the use of data analytics in DoD, including the establishment of an Office of People Analytics (OPA) to better harness DoD's big data capabilities in managing its talent. While few details have been released at the time of this report's publication, OPA is designed to provide direct analytic support to the Services and OSD. Such support would give researchers and analysts better access to data on personnel characteristics and would allow them to conduct comprehensive analyses on how policy or environmental changes will affect the performance or composition of the workforce. OPA will be prepared to partner with the Services and OSD on questions pertaining to recruiting, hiring and retention, succession planning, and training and would improve their ability to match the talents of individual personnel to the talents demanded by the jobs they are assigned to.

Improved data and descriptive, predictive, and prescriptive analytics, which are discussed in generality in Chapter 4, could be more effectively used to explore and evaluate policy options. Following are the six principal outcomes for which P&R is responsible (introduced in Chapter 2) and the committee's observations about how advanced data analytics could help strengthen the way they are addressed:

- *Ensuring DoD can recruit, train, motivate, and retain the necessary numbers and qualities of personnel.* Improved access to and sharing of data, and possibly the use of novel data sources, allows for a deeper understanding of how best to recruit, train, motivate, and retain the necessary numbers and qualities of personnel. The private sector faces many similar personnel challenges and has developed and deployed descriptive, predictive, and prescriptive analytics to address these workforce issues. There are unique considerations for DoD, but understanding the large-scale efforts in the private sector is a starting point. Important workforce issues can then be examined through variations of these methods, coupled with DoD data.

  This report's recommendations to establish a process for identifying domains in which data sharing would be of biggest benefit; develop a working group from the Services and other elements of DoD to improve coordination across groups; and leverage opportunities to improve and incentivize data sharing would all help this P&R area.
- *Creating incentives that guide DoD to an "optimal" mix of personnel.* Increased use of prescriptive analytics could help inform staffing-level decisions for DoD personnel. This might include the various methods associated with both stochastic models (e.g., analytical, numerical, and simulation methods) and mathematical optimization (e.g., stochastic optimization and stochastic optimal control methods) to study options and determine the best decisions.

  Controlled experiments such as small-scale pilot projects could also help inform policy decisions relating to this area (e.g., the effectiveness of incentives). The outcome of these controlled experiments could then be used as input into the prescriptive analytics models and methods, thus improving the analytics for future use.
- *Ensuring DoD creates a force that is ready to carry out directed*

actions. Readiness data can be explored using improved data analytics, therefore making it easier to detect underprepared units or capabilities and to estimate the impact of any deficiencies. This has the potential to target available resources more effectively to gain greater benefit from investments.

Having a strong data science capability for DoD could also be considered its own aspect of readiness. Since many data analytics advances are being employed in the private sector, it is reasonable to assume that they could also be developed within other military forces. The task of ensuring DoD maintains a force that is ready to carry out directed actions, even with a shrinking budget and growing demand for talent, could be helped by improved analytics.

- *Influencing DoD's decisions that affect the shape of military careers*. Improved data access and sharing and improved analytics capabilities could significantly improve the knowledge base for shaping military careers. Better understanding the range of career paths and which positions tend to lead to certain outcomes would be of value to personnel making career choices and to leaders evaluating whether talent is being used well. The private sector has made considerable investments in understanding and shaping career paths, and DoD could benefit from exploring those efforts.
- *Ensuring the services supporting DoD personnel are properly structured and provided*. Monitoring the services supporting DoD personnel is a data-intensive challenge, and so steps to improve access to data and effective use of data analytics would be helpful. Increased use of prescriptive analytics, including the various methods associated with both stochastic models (e.g., analytical, numerical, and simulation methods) and mathematical optimization (e.g., stochastic optimization and stochastic optimal control methods) in particular could help assess if the services are properly structured and provided.
- *Anticipating and responding to sensitive behavioral issues*. Data analytics have the potential to help identify individuals at risk for sensitive behavioral issues and ensure that responses are proportionate and consistent. Improved understanding could be enabled both by improved access to existing data as well as to new sources of data. However, use of these data and methods requires careful consideration because of privacy implications, and data analytics must be carried out with care and in compliance with appropriate federal guidelines.

While the preceding examples illustrate some ways in which improved data analytic approaches could assist P&R with its mission, these are only the beginning of what could be possible. As can be seen, some particularly striking opportunities could follow as capabilities for predictive and prescriptive analytics are built up. As discussed in Chapter 6, the committee does not believe current personnel analytics tools are a good match for DoD needs in part because they are largely targeted at operational decision making related to HR and aimed at a targeted market segment that does not match DoD. P&R may gain the most benefit from developing tools guided by P&R's best understanding of the relative priorities of the various decisions within its mission. Many of P&R's individual challenges are being addressed using some aspects of data analytics, but the collection of those individual studies falls short of a cohesive plan of how and why to use these approaches across the many domains that P&R informs.

> **Finding:** A wide range of problems are being addressed for P&R using data analytic techniques and the rich data sources discussed in this report. These are often applied in response to specific questions but are not incorporated into a long-term plan.
>
> **Finding:** Turnkey personnel analytic solutions and currently commercially available software are unlikely to meet P&R's needs.
>
> **Recommendation 11:** The Office of the Under Secretary of Defense (Personnel & Readiness) should assess which predictive and prescriptive analyses would benefit its mission over the longer term, taking into account its understanding of which specific decisions could, if evaluated by applying more powerful data and/or methods, better enable the Department of Defense to prepare for future demands it may face. Some possible steps that might follow include these:
>
> - Emphasizing the use of prescriptive analytics in conjunction with predictive what-if scenarios;
> - Enhancing prescriptive analytics usage and disseminating best practices across the entire department; and
> - Adapting the prescriptive analytics methods successfully used in the private sector for workforce and talent management.

Controlled Experiments

Often P&R needs additional data to evaluate potential policy solutions, and the use of controlled experiments can sometimes provide those data. Controlled experiments are an opportunity for P&R to test a hypothesis by adjusting a key variable in a subset of its population and then measuring the results of the change. Besides the challenge of structuring the experiment, the resources required and the time delay in obtaining results can inhibit the use of controlled experiments. The urgency of solving a problem frequently leads to deploying solutions without a framework for testing their efficacy except through pre- and post-solution data analysis. The result is an institutional culture in which experiments are the exception rather than the rule (as opposed to, say, the civil health community); however, this is a missed opportunity.

Controlled experiments can be used for a variety of areas important to P&R (e.g., recruiting [Fricker et al., 2015]) and could be particularly helpful in buttressing the case for conclusions that contradict accepted propositions. A case in point is the enthusiasm for 2-year enlistments, endorsed by some political leaders as a means of shifting the supply of enlistees (particularly those from socioeconomic backgrounds that would not otherwise normally lead to enlisted military service). A controlled Army experiment demonstrated that any enlistment supply gain would be swamped by the attrition at the end of the service period while imposing unnecessary training costs engendered by high turnover and a short period of trained service within the 2-year period. Most skill areas require 6 to 12 months of training before assignment to a unit, at which point the individual is still at the initial point of the learning curve (Buddin, 1991).

The NRC (2004) reviewed a variety of experiments and quasi-experiments (in addition to the Buddin [1991] study described above) that examined the effects of various recruiting incentives on enlistment, job choice, and other outcomes. For example, Fernandez (1982) reported a study of the effects on high-quality enlistment of differing types and amounts of postservice educational benefits. Polich et al. (1986) examined the effects of varying enlistment bonuses and differing enlistment term obligations on highly qualified enlistments. Sellman (1999) reported on a pilot program allowing recruits to attend 2 years of college prior to reporting for active duty.

Finding: The Department of Defense does not routinely employ controlled experiments to understand causes and effects of the Office of the Under Secretary of Defense (Personnel & Readiness) policies—for example, revisions to enlistment standards or choices affecting family welfare—to judge whether they produce the intended effects and provide benefits that justify their costs.

Recommendation 12: To the extent feasible and relevant, the Department of Defense should conduct carefully structured experiments to test the efficacy of policy.

STRENGTHEN P&R WORKFORCE CAPABILITIES IN DATA SCIENCE

The demand for individuals with data science skills is increasing rapidly across all organizations (inside and outside of DoD). For example, the U.S. Bureau of Labor Statistics reports that the expected rate of growth for statisticians is 34 percent (BLS, 2016a) and for operations research analysts (a kind of data science analyst) is 30 percent (BLS, 2016b) over the period 2014-2024 as compared with the average growth rate for all occupations of 7 percent. *U.S. News and World Report* ranks "operations research analyst" as the fourth best business job, the eighth best STEM job, and the twentieth best job overall by incorporating factors such as job growth and salary (Marquardt, 2015). Other kinds of surveys and rankings provide similar results. Individuals with data science skills that include descriptive, predictive, and prescriptive analytics experience are in strong demand for employment and salary growth. The private sector is fueling this growth; in response, more than 80 university programs have been created over the past 10 years to provide undergraduate and graduate analytics education (INFORMS, 2015).

There are significant opportunities within P&R for increased use of data science to improve a variety of aspects of decision making. It is notable that DoD employs very few individuals with expertise in statistics and optimization, and many quantitatively trained analysts are classified as operations researchers, regardless of their actual training.

Finding: Based on its collective experience with seeing data science mature in other organizations, the committee's judgment is that P&R's skills, depth, and resources in data analytics are not sufficient to recognize the full range of analytics opportunities and to implement these methods to better support decision making. It is always problematic to leverage scattered pockets of data science expertise, so raising the general level of awareness and skill would be more effective.

Recommendation 13: The Office of the Under Secretary of Defense (Personnel & Readiness) should create greater awareness of data science methods and disseminate them more thoroughly to its personnel to increase the general understanding of data science and the benefits of its use.

Recommendation 14: The Office of the Under Secretary of Defense (Personnel & Readiness) should enhance education in data science for its personnel, including civil service employees. This education could range from short courses in specific techniques for personnel who already have the requisite foundational knowledge, to overview seminars for managers who need to be acquainted with what their analytical staff can undertake, to formal degree programs, whether at Department of Defense or civilian universities.

CONCLUSIONS

The heart of P&R's responsibility is policy management for DoD personnel and assessment of their readiness to carry out the tasks the nation assigns. This mission requires high-quality data and the most up-to-date means to analyze it. High-quality, timely data are necessary to understand events now occurring (or that have occurred in the recent past) and to forecast what may happen next, to which policies must be ready to respond, if not to take preemptive action to thwart adverse outcomes. High-quality data are necessary to understand the structure of those events (that is to say, the causal factors), lest the policy choices deal with the symptoms observed versus the underlying causes.

P&R today commands an extraordinary variety of data sets and data sources when compared to what is available to most cabinet agencies. But much of what is available is based on administrative records or records that are driven in key ways by administrative considerations. What is available can be shaped into data sets that respond to P&R's policy analysis needs, albeit at some cost in resources and the degree to which the information is available on a timely basis. If P&R desires more timely information, and information that is more complete relative to its needs—including the ability to forecast and evaluate alternative policy decisions, which good policy debate requires—it will need to consider additional investments. And it will need to devise mechanisms for controlling data access that on the one hand protect the variety of equities involved (including privacy), but on the other respond more quickly and adequately to the analytic needs that its own requests have often generated. Secretary Carter's Force of the Future recommends the creation of an Office of People Analytics. This report is intended to provide a way to confront these issues in a manner that will bring P&R, DoD, and the American military into a commanding position of excellence in managing personnel.

REFERENCES

Bicksler, B.A., C.L. Gilroy, and J.T. Warner, eds. 2004. *The All-Volunteer Force: Thirty Years of Service*. Dulles, Va.: Brassey's.

BLS (Bureau of Labor Statistics). 2016a. U.S. Department of Labor. Occupational Outlook Handbook. 2016-17 Edition, Operations Research Analysts. http://www.bls.gov/ooh/math/statisticians.htm. Accessed April 8, 2016.

BLS. 2016b. Occupational Outlook Handbook. 2016-17 Edition, Operations Research Analysts. http://www.bls.gov/ooh/math/operations-research-analysts.htm. Accessed April 8, 2016.

Bowman, W., R. Little, and G.T. Sicilia, eds. 1986. *The All-Volunteer Force After a Decade: Retrospect and Prospect*. McLean, Va.: Pergamon-Brassey's.

Buddin, R. 1991. *Enlistment Effects of the 2 + 2 + 4 Recruiting Experiment*. Santa Monica, Calif.: RAND Corporation. http://www.rand.org/pubs/reports/R4097.

Calandrino, J.A., A. Kilzer, A. Narayanan, E.W. Felten, and V. Shmatikov. 2011. "You Might Also Like": Privacy risks of collaborative filtering. Pp. 231-246 in *Proceedings of the 2011 IEEE Symposium on Security and Privacy*. May 22-25.

Carter, A. 2015. Remarks on "Building the First Link to the Force of the Future" as Delivered by Secretary of Defense Ash Carter. George Washington University Elliott School of International Affairs, Washington, D.C., November 18. http://www.defense.gov/News/Speeches/Speech-View/Article/630415/remarks-on-building-the-first-link-to-the-force-of-the-future-george-washington.

Fernandez, R. 1982. *Enlistment Effects and Policy Implications of the Educational Test Assistance Program*. Report R-2935-MRAL. Santa Monica, Calif.: RAND Corporation.

Fredland, J.E., C. Gilroy, R.D. Little, and W.S. Sellman, eds. 1996. *Professionals on the Front Line: Two Decades of the All-Volunteer Force*. Washington, D.C.: Brassey's.

Fricker, R.D., Jr., S.E. Buttrey, and J.K. Alt. 2015. Future Navy Recruiting Strategies. *Naval Postgraduate School*. http://faculty.nps.edu/rdfricke/docs/CNRC_technical_report.pdf.

Ganta, S.R., S. Kasiviswanathan, and A. Smith. 2008. Composition attacks and auxiliary information in data privacy. In *Proceedings of the 14th ACM SIGKDD International Conference on Knowledge Discovery and Data Mining*. http://www.cse.psu.edu/~ads22/privacy598/papers/gks08.pdf.

INFORMS (Institute for Operations Research and the Management Sciences). 2015. "Analytics and OR/MS Education." https://education.informs.org. Accessed November 17, 2015.

Losey, S. 2016. "Need to Know, 2016: Air Force Offers Three-Year Sabbaticals." *Air Force Times*. January 1. http://www.airforcetimes.com/story/military/careers/air-force/2016/01/01/need-know-2016-air-force-offers-three-year-sabbaticals/77762608/.

Marquardt, K. 2015. "Best Business Jobs: Operations Research Analysts." Overview. *U.S. News and World Report*. http://money.usnews.com/careers/best-jobs/operations-research-analyst.

Narayanan, A., and V. Shmatikov. 2008. Robust de-anonymization of large sparse datasets. Pp. 111-125 in *Proceedings of the 2008 IEEE Symposium on Security and Privacy*. doi: 10.1109/SP.2008.33.

NRC (National Research Council). 2004. *Evaluating Military Advertising and Recruiting: Theory and Methodology*. Washington, D.C.: The National Academies Press.

NRC. 2014. *Proposed Revisions to the Common Rule for the Protection of Human Subjects in the Behavioral and Social Sciences*. Washington, D.C.: The National Academies Press.

Polich, M., J. Dertouzos, and J. Press. 1986. The Enlistment Bonus Experiment. Report No. R-3353-FMP. Santa Monica, Calif.: RAND Corporation.

Schechter, E. 2016. "Defense Secretary Outlines Strategies, Goals." *Pensacola News Journal*. January 30. http://www.pnj.com/story/news/military/2016/01/30/defense-secretary-outlines-strategies-goals/79536406/.

Sellman, W.S. 1999. Military Recruiting: The Ethics of Science in a Practical World. Invited address to the Division of Military Psychology, 107th annual convention of the American Psychological Association, Boston, Mass.

Serbu, J. 2016. "DoD Bids to Make Military Life More 'Family-Friendly.'" *Federal News Radio*. January 29. http://federalnewsradio.com/defense/2016/01/dod-doubles-maternity-leave-bid-make-military-life-family-friendly/.

Stark, S., O.S. Chernyshenko, F. Drasgow, C.D. Nye, W.L. Farmer, L.A. White, and T. Heffner. 2014. From ABLE to TAPAS: A new generation of personality tests to support military selection and classification decisions. *Military Psychology* 26:153-164.

Sweeney, L. 1997. Weaving technology and policy together to maintain confidentiality. *Journal of Law, Medicine and Ethics* 25(2-3):98-110.

Tene, O., and J. Polonetsky. 2014. A theory of creepy: Technology, privacy, and shifting social norms. *Yale Journal of Law and Technology* 16(1):2.

Wilmoth, J.M., and A.S. London. 2013. *Life-Course Perspectives on Military Service*. New York: Routledge.

Appendixes

A

Acronyms

ACES	Automated Continuous Evaluation System
ASD	Assistant Secretary of Defense
ASVAB	Armed Services Vocational Aptitude Battery
CAC	Common Access Card
CART	Classification and Regression Tree
CIPSEA	Confidential Information Protection and Statistical Efficiency Act
CNA	Center for Naval Analyses
DASD	Deputy Assistant Secretary of Defense
DBIDS	Defense Biometric Identification System
DBMS	Database Management System
DEERS	Defense Enrollment Eligibility Reporting System
DEP	Delayed Entry Program
DHRA	Defense Human Resources Activity
DIBRS	Defense Incident-Based Reporting System
DIOR	Directorate of Information Operations and Reports
DMDC	Defense Manpower Data Center
DoD	Department of Defense
DRRS	Defense Readiness Reporting System
DTIC	Defense Technical Information Center
ETL	extract-transform-load

FFRDC	federally funded research and development center
FIPPs	Fair Information Practice Principles
G/FO	General and Flag Officer
GSA	General Services Administration
HIPAA	Health Insurance Portability and Accountability Act
HR	human resources
HumRRO	Human Resources Research Organization
IDA	Institute for Defense Analyses
INFORMS	Institute for Operations Research and the Management Sciences
IRB	Institutional Review Board
JAMS	Joint Adjudication Management System
JCAVS	Joint Clearance and Access Verification System
JDAMIS	Joint Duty Assignment Management Information System
JPAS	Joint Personnel Adjudication System
M&RA	Manpower and Reserve Affairs
MEPS	military entrance processing station
MHS	Military Health Service
MS&A	modeling, simulation, and analysis
NAE	National Academy of Engineering
NAS	National Academy of Sciences
NCHS	National Center for Health Statistics
NDRI	National Defense Research Institute
NER	named entity recognition
NIST	National Institute of Standards and Technology
NLP	natural language processing
NLSY79	National Longitudinal Survey of Youth 1979
NPS	Naval Postgraduate School
NRC	National Research Council
ODMEO	Office of Diversity Management and Equal Opportunity
OMB	Office of Management and Budget
OPA	Office of People Analytics
OPM	Office of Personnel Management
OSD	Office of the Secretary of Defense

P&R	Office of the Under Secretary of Defense (Personnel & Readiness)
PCAST	President's Council of Advisors on Science and Technology
PDASD	Principal Deputy Assistant Secretary of Defense
PDE	Person-Event Data Environment
PDHA	Post-Deployment Health Assessment
PDHRA	Post-Deployment Health Re-Assessment
PDUSD	Principal Deputy Under Secretary of Defense
PII	personally identifiable information
PTS	Post-Traumatic Stress
SAPRO	Sexual Assault Prevention and Response Office
SORTS	Status of Resources and Training System
SVM	support vector machine
TAPAS	Tailored Adaptive Personality Assessment System
VA	Department of Veterans Affairs
YRRP	Yellow Ribbon Reintegration Program

B

Biographies of the Committee

STEPHEN M. ROBINSON, *Chair*, is professor emeritus in the Department of Industrial and Systems Engineering at the University of Wisconsin, Madison. In 2008, he was elected as a member of the National Academy of Engineering (NAE), Section 8 - Industrial, Manufacturing, and Operational Systems Engineering. His research is in the development of quantitative methods for making the best use of scarce resources, which is part of the broad category of operations research methods. He works particularly on nonlinear and stochastic optimization methods for both optimization and equilibrium problems, trying both to develop the underlying theory and to find better numerical methods for solving applied problems. His recent work has focused especially on the mathematical properties of solutions of variational condition, considered as functions of the data appearing in those conditions. He has published more than 100 articles and recently finished a term as president of INFORMS. Dr. Robinson received his Ph.D. in computer sciences from the University of Wisconsin, Madison, in 1971.

PASCALE CARAYON is Procter & Gamble Bascom Professor in Total Quality in the Department of Industrial and Systems Engineering and the Director of the Center for Quality and Productivity Improvement (CQPI) at the University of Wisconsin, Madison. She leads the Systems Engineering Initiative for Patient Safety (SEIPS) at the University of Wisconsin, Madison. She received her engineering diploma from the École Centrale de Paris, France, in 1984, and her Ph.D. in industrial engineering from the University of Wisconsin, Madison, in 1988. Her research in human factors and ergonomics focuses on macroergonomics and aims at modeling, assessing, and improving sociotechnical systems in order to enhance health care quality and patient safety. This research has been funded by the Agency for Healthcare Research and Quality, the National Science Foundation (NSF), the National Institutes of Health (NIH), the National Institute for Occupational Safety and Health, the Department of Defense (DoD), the office of the National Coordinator for Health Information Technology, various foundations, and private industry. She is the co-editor-in-chief for *Applied Ergonomics* and a member of the editorial boards of the *Journal of Patient Safety, Behaviour and Information Technology*, and *Work and Stress*. She is a fellow of the Human Factors and Ergonomics Society (HFES) and the International Ergonomics Association (IEA). She is the recipient of the IEA Triennial Distinguished Service Award (2012). Dr. Carayon is a member of the HFES executive council. She is the editor of the *Handbook of Human Factors and Ergonomics in Health Care and Patient Safety*.

DAVID CHU serves as president of the Institute for Defense Analyses (IDA), a nonprofit corporation operating in the public interest. Its three federally funded research and development centers provide objective analyses of national security issues and related national challenges, particularly those requiring extraordinary scientific and technical expertise. As president, Dr. Chu directs the activities of more than 1,000 scientists and technologists. Together, they conduct and support research requested by federal agencies involved in advancing national security and advising on science and technology issues. Dr. Chu served in the DoD as Under Secretary of Defense (Personnel & Readiness) from 2001 to 2009, and earlier as Assistant Secretary of Defense and Director for Program Analysis and Evaluation from 1981 to 1993. From 1978 to 1981 he was the assistant director of the Congressional Budget Office for National Security and International Affairs. Dr. Chu served in the U.S. Army from 1968 to 1970. He was an economist with the RAND Corporation from 1970 to 1978, director of RAND's Washington office from 1994 to 1998, and vice president for its Army Research Division from 1998 to 2001. He earned a bachelor of arts in economics and mathematics and his doctorate in economics from Yale University. Dr. Chu is a member of the Defense Science Board and a fellow of the National Academy of Public Administration. He is a recipient of the DoD Medal for Distinguished Public Service with Gold Palm, the Department of Veterans Affairs Meritorious Service Award, the Department of the Army Distinguished Civilian Service Award, the Department of the Navy Distinguished Public Service Award, and the National Academy of Public Administration's National Public Service Award.

CYNTHIA DWORK is a Distinguished Scientist at Microsoft Research. Dr. Dwork has made seminal contributions to cryptography and distributed computing. She invented differential privacy (with McSherry, Nissim, and Smith), cryptosystems whose security is based on worst-case analysis (with Ajtai), nonmalleable cryptography (with Dolev and Naor), and computational-effort-based antispam techniques (with Naor). She is a member of the National Academy of Sciences (NAS) and the NAE, and she was elected as a fellow of the American Academy of Arts and Sciences (AAAS) in 2008. She received the Dijkstra Prize in 2007 for her work on consensus problems together with Nancy Lynch and Larry Stockmeyer. In 2009 she won the PET Award for Outstanding Research in Privacy Enhancing Technologies. Dr. Dwork received her B.S.E. from Princeton University in 1979, graduating cum laude, and receiving the Charles Ira Young Award for Excellence in Independent Research. She received her Ph.D. from Cornell University in 1983.

TERRY P. HARRISON is a professor of supply chain and information systems, the Earl P. Strong Professor in Business, and the director of business analytics in the Smeal College of Business at Pennsylvania State University. His teaching and research interests are in the areas of supply chain management and modeling, large-scale production and distribution systems, decision support systems, and applied optimization. He received a B.S. in forest science/forest products from Pennsylvania State University and an M.S. and Ph.D. in management science from the University of Tennessee. He is a past president and fellow of INFORMS, former vice president of publications for INFORMS, and former editor-in-chief of *Interfaces*. He has been recognized as an Edelman laureate and as a Wilbur B. Payne award recipient for excellence in analysis by the U.S. Army.

ALAN F. KARR is director of the Center of Excellence for Complex Data Analysis (CoDA) at RTI International. Dr. Karr is the former director of the National Institute of Statistical Sciences, located in Research Triangle Park, North Carolina. He was director for 14 years and previously was assistant director from 1992 to 2000. He is also a professor of statistics and biostatistics at the University of North Carolina, Chapel Hill. He received his bachelor of science degree in industrial engineering from Northwestern University in 1969 with highest distinction. He continued at Northwestern University, receiving his masters of science degree in industrial engineering in 1970 and his Ph.D. in applied mathematics in 1973. Dr. Karr spent roughly 20 years at Johns Hopkins University in the Mathematical Sciences Department, serving as chair in 1985 and 1986. He also served as associate dean, G.W.C. Whiting School of Engineering, from 1986 to 1992. Dr. Karr has published over 110 scientific papers and written two books, *Point Processes and Their Statistical Inference* and *Probability*. He is a contributing editor for the *IMS Bulletin*. He has also served on the Army Science Board and has been an associate editor for the *Operations Research Letters*, *Mathematics of Operations Research*, and the *SIAM Journal on Applied Mathematics*. He is also a fellow of the ASA and the Institute of Mathematical Sciences, and an elected member of the International Statistical Institute.

SALLIE KELLER, Ph.D., is professor of statistics and director of the Social and Decision Analytics Laboratory within the Virginia Bioinformatics Institute at Virginia Tech. Formerly, she was professor of statistics and academic vice president and provost at University of Waterloo, director of the IDA Science and Technology Policy Institute, and professor of statistics and the William and Stephanie Sick Dean of Engineering at Rice University. Her other appointments include head of the Statistical Sciences group at Los Alamos National Laboratory, professor and director of graduate studies in the Department of Statistics at Kansas State University, and statistics program director at NSF. For the National Academies of Sciences, Engineering, and Medicine, Dr. Keller has served as a member of the Board on Mathematical Sciences and Their Applications, has chaired the Committee on Applied and Theoretical Statistics, and is currently a member of the Committee on National Statistics. Her areas of expertise are social and decision informatics, the statistical underpinnings of data science, uncertainty quantification, and data access and confidentiality. She is a national associate of the NAS, fellow of the American Association for the Advancement of Science, elected member of the International Statistics Institute, and member of the JASON advisory group. She is also a fellow and past president of the American Statistical Association. She holds a Ph.D. in statistics from the Iowa State University of Science and Technology.

ALAIR MACLEAN is associate professor of sociology at Washington State University, Vancouver. Her research focuses broadly on social inequality. She is currently exploring the life course trajectories of veterans who served in the U.S. armed forces, focusing on the effects of military service and combat exposure on work and health. She has published articles in the *Annual Review of Sociology*, *American Review of Sociology*, and *Sociology of Education*. She is a member of the American Sociological Association, the Population Association of America, and the Inter-University Seminar on Armed Forces and Society. She is currently a member of the National Academies' Committee on the Initial Assessment of Readjustment Needs of Military Personnel, Veterans and Their Families. She received an M.S. and a Ph.D. in sociology from the University of Wisconsin, Madison, and completed a postdoctoral fellowship at the RAND Corporation.

DAVID MAIER is the Maseeh Professor of Emerging Technologies at Portland State University. Prior to his current position, he was on the faculty at the State University of New York (SUNY)–Stony Brook and Oregon Graduate Institute. He has spent extended visits at the Institut National de Recherche en Informatique et en Automatique, University of Wisconsin, Madison, Microsoft Research, and the National University of Singapore. He is the author of books on relational databases, logic programming, and object-oriented databases, as well as papers in database theory, object-oriented technology, scientific databases, and dataspace management. He is a recognized expert on the challenges of large-scale data in the sciences. He received an NSF Young Investigator Award in 1984 and was awarded the 1997 SIGMOD Innovations Award for his contributions in objects and databases. He is an Association for Computing Machinery (ACM) fellow and Institute of Electrical and Electronics Engineers (IEEE) senior member. He holds a dual B.A. in mathematics and in computer science from the University of Oregon (Honors College) and a Ph.D. in electrical engineering and computer science from Princeton University.

PAUL R. SACKETT is the Beverly and Richard Fink Distinguished Professor of Psychology and Liberal Arts at the University of Minnesota. His research interests revolve around various aspects of testing and assessment in workplace, military, and educational settings. His work on issues of fairness and bias in testing includes frequently cited 1994, 2001, and 2008 *American Psychologist* articles. He has long been active in the area of the assessment of honesty and integrity in the workplace. He also publishes extensively on the assessment of managerial potential and methodological issues in employee selection. He has worked with a wide variety of public- and private-sector organizations on the design and evaluation of selection and training systems. He served as founding editor of the Society for Industrial and Organizational Psychology's (SIOP's) journal *Industrial and Organizational Psychology: Perspectives on Science and Practice* and editor of *Personnel Psychology*. He has served as president of SIOP, as cochair of the Joint Committee on the Standards for Educational and Psychological Testing, as a member of the National Academies' Board on Testing and Assessment, as chair of the American Psychological Association's (APA's) Committee on Psychological Tests and Assessments, and as chair of APA's Board of Scientific Affairs. He received his Ph.D. in industrial and organizational psychology at the Ohio State University.

MARK S. SQUILLANTE is a distinguished research staff member and the area head of stochastic processes, optimization, and control within the Mathematical Sciences Department at the IBM Thomas J. Watson Research Center. He serves as director of the Center for Optimization under Uncertainty Research across IBM Research. He has been an adjunct faculty member at Columbia University and a member of the technical staff at Bell Laboratories, and has held visiting positions at various academic institutions. His research interests concern mathematical foundations of the analysis, modeling, and optimization of the design and control of complex stochastic systems, with applications across a broad spectrum of areas including decision making under uncertainty, business analytics and optimization, data analytics, smarter energy/planet/workforce technologies, health analytics, and social media analytics. He is an elected fellow of ACM and IEEE and the author of more than 250 technical papers and more than 30 issued or filed patents. His work has been recognized through the Daniel H. Wagner Prize (INFORMS), 8 best paper awards, 11 keynote/plenary presentations, 14 major IBM technical awards, and 26 IBM invention awards. He is a member of AMS, Bernoulli Society, IMS, INFORMS, and SIAM. He serves on the editorial boards of *ACM Transactions on Modeling and Performance Evaluation of Computing Systems*, *Performance Evaluation*, and *Stochastic Models*, is co-editor of *Matrix-Analytic Methods in Stochastic Models*, and has chaired/organized several international conferences in applied probability and related applications. He received a Ph.D. from the University of Washington.

WILLIAM J. STRICKLAND is president and chief executive officer of the Human Resources Research Organization (HumRRO) in Alexandria, Virginia. Before his appointment as CEO, he spent over 10 years as a HumRRO vice president, directing its Workforce Analysis and Training Systems Division. Before joining HumRRO, he served in the United States Air Force and retired with the rank of colonel; in his last assignment, he was the director for Air Force human resources research. He is a fellow of the APA and past president of its Division of Military Psychology and served for 6 years as that division's representative on the APA Council of Representatives. He currently serves as a member-at-large on the APA board of directors and has represented APA on the Board of the Consortium of Social Science Organizations and on the Council of the Federation for Brain and Behavioral Sciences. He has been a member of four previous National Academies committees. He is a graduate of the United States Air Force Academy and earned a Ph.D. in industrial and organizational psychology from Ohio State University.

STEVEN TADELIS is an associate professor of business and public policy at the Haas School of Business at the University of California, Berkeley. His research primarily revolves around e-commerce and the economics of the Internet. During the 2011-2013 academic years he was on leave at eBay research labs, where he hired and led a team of research economists. Their work focused on the economics of e-commerce, with particular attention to creating better matches of buyers and sellers, reducing market frictions by increasing trust and safety in eBay's marketplace, understanding the underlying value of different advertising and marketing strategies, and exploring the market benefits of different pricing structures. Aside from the economics of e-commerce, his main fields of interest are the economics of incentives and organizations, industrial organization, and microeconomics. Some of his past research aspired to advance our understanding of the roles played by two central institutions—firms and contractual agreements—and how these institutions facilitate the creation of value. Within this broader framework, he has explored firm reputation as a valuable, tradable asset; the effects of contract design and organizational form on firm behavior with applications to outsourcing and privatization; public- and private-sector procurement and award mechanisms; and the determinants of trust. He received a Ph.D. in economics from Harvard University, an M.Sc. in economics from the Technion in Israel, and a B.A. in economics from the University of Haifa in Israel.

C

Meetings and Presentations

FIRST COMMITTEE MEETING
MARCH 30-31, 2015

Origin and Importance of the Study
Mark Breckenridge, Deputy Director, Defense Manpower Data Center

Study Impact for DoD
Christine Fox, Assistant Director for Policy and Analysis, Johns Hopkins University Applied Physics Laboratory

RAND Overview of Data and Analytics of Value to P&R
John D. Winkler, Director of the Forces and Resources Policy Center, RAND National Defense Research Institute

CNA Overview of Data and Analytics of Value to P&R
Linda C. Cavalluzzo, Vice President and Director of Resource Analysis, Center for Naval Analysis

Motivation for the Defense Technical Information Center (DTIC) text analytics study
Mona Lush, Special Assistant for Acquisition Initiatives, Office of the Under Secretary of Defense (Acquisition, Technology & Logistics)/Acquisition Resources and Analysis

Overview of the DTIC text analytics study
Laura Odell, Project Leader for DTIC study, Institute for Defense Analyses
Arun Maiya, Principal Investigator for DTIC study, Institute for Defense Analyses

SECOND COMMITTEE MEETING
MAY 4-5, 2015

Talent Analytics with IBM Smarter Workforce
Jackie Ryan, Director, Science and Analytics Project Management, IBM Smarter Workforce
Colum Foley, IBM Smarter Workforce

Office of Personnel & Readiness
Stephanie Barna, Acting Assistant Secretary of Defense for Readiness and Force Management

Defense Health Agency
Rear Adm. Bruce Doll, Director, Research, Development and Acquisition, Defense Health Agency

Challenges Facing the Civilian Workforce
Christine Fox, Assistant Director for Policy and Analysis, Johns Hopkins University Applied Physics Laboratory

THIRD COMMITTEE MEETING
OCTOBER 6-7, 2015

Site Visit Briefings: Defense Readiness Reporting System, Navy, Intel, Army, Air Force, P&R, RAND, CNA, IDA, NPS, DMDC, Google, Cornerstone OnDemand, and Workday
Michelle Schwalbe, Senior Program Officer, Board on Mathematical Sciences and Their Applications

FOURTH COMMITTEE MEETING
JANUARY 6, 2016

Discussion of the Person-Event Data Environment
Mark Breckenridge, Defense Manpower Data Center
Matthew Davis, Defense Manpower Data Center

D

Stochastic Models of Uncertainty and Mathematical Optimization Under Uncertainty

This appendix provides more formal definitions and descriptions of aspects of the two key areas of prescriptive analytics, namely stochastic models of uncertainty and mathematical optimization under uncertainty, which are intimately connected. These descriptions also illustrate examples of mathematical prescriptive analytics methods that have been developed and successfully applied to address policy issues and problems associated with workforce management, skill and talent management, and human capital resource allocation in the private sector, which are analogous to the type of prescriptive decision-making policy issues and problems related to missions of the Office of the Under Secretary of Defense (Personnel & Readiness), referred to as P&R, including recruiting, training, retention, optimal mix, readiness, staffing, career shaping, and resource allocation. Although there are unique considerations for the Department of Defense (DoD), an understanding of these and related large-scale efforts in the private sector is an important starting point. Then policy issues and problems related to P&R missions can be investigated through variations of such methods, coupled with DoD data.

The following descriptions are highly mathematical, relative to the rest of this report, and will likely be relevant to only a subset of readers. However, the committee believes these examples may be helpful to analysts trying to implement these methods.

STOCHASTIC MODELS OF UNCERTAINTY

To illustrate some of the concepts described in Chapter 4, two examples of stochastic models of uncertainty involved in decision-making problems related to P&R are presented. The first example concerns trade-offs among skill capacities and readiness of resources given uncertainty around the demand for such resources, which relate to P&R missions associated with optimal mix, readiness, staffing, and resource allocation. This material is based on the works of Bhadra et al. (2007), Cao et al. (2011), Jung et al. (2008), Jung et al. (2014), and Lu et al. (2007), where additional technical details can be found. The second example concerns trade-offs among the evolution of capabilities and readiness of personnel resources over time given uncertainty around time-varying supply-side dynamics, which relate to P&R missions associated with recruiting, training, retention, optimal mix, readiness, staffing, career shaping, and resource allocation. This material is based on the work of Cao et al. (2011), where additional technical details can be found.

Example: Stochastic Loss Networks

One such example is based on the use of stochastic loss networks to model the probability of sufficient capacity and readiness of resources given uncertainty around the demand for such resources, as well as the interactions and dynamics of resources across different projects and tasks. As a specific illustrative instance related to P&R, analogous to the workforce application studied in Cao et al. (2011), consider the problem of determining the likelihood of not having enough resources of type $i \in I$ (e.g., experienced aircraft mechanics) in order to satisfy the demand for projects of type $k \in K$ (e.g., sustaining high aircraft mission-capable rates), where I denotes the set of resource types and K denotes the set of project types. Let $A_{i,k}$ denote the amount of type-i resources that are required to deliver type-k projects and C_i the total amount of type-i resources available. The demand for type-k projects is modeled as a stochastic arrival process with rate v_k (derived from the output of predictive analytics)—that is, the times between the need to deliver type-k projects follow a probability distribution with mean v_k^{-1}. When delivery of a type-k project is needed and the amount of type-i resources available for deployment at that time is at least $A_{i,k}$ for all $i \in I$, then the type-k project is delivered and the required resources of each type are deployed for a duration that follows an independent probability distribution with unit mean (without loss of generality); otherwise, the type-k project cannot be delivered and the opportunity to do so is said to be lost. Define L_k to be the stationary loss-risk probability for type-k projects and n_k the number of active type-k projects in equilibrium, with $\boldsymbol{n} := (n_k)$. By definition, we have $\boldsymbol{n}$ as an element of the space of possible deployments

$$S(\mathrm{C}) := \left\{ \boldsymbol{n} \in \mathbb{Z}_+^{|K|} : \boldsymbol{A}\boldsymbol{n} \leq \mathrm{C} \right\},$$

where $\boldsymbol{A} := \left[A_{i,k}\right]$, $\mathrm{C} := (C_i)$, and $\mathbb{Z}_+$ denotes the set of non-negative integers. Assuming independent Poisson arrival processes with rate vector $\boldsymbol{v} = (v_k)$, the key probability measure of interest then can be expressed as

$$L_k = 1 - G(\mathrm{C})^{-1}\, G(\mathrm{C} - Ae_k), \tag{A.1}$$

$$G(\mathrm{C}) = \sum_{\boldsymbol{n} \in S(\mathrm{C})} \prod_{k \in K} \frac{v_k^{n_k}}{n_k!}, \tag{A.2}$$

where e_k is the unit vector corresponding to a single active type-k project. Refer to Jung et al. (2008) and Jung et al. (2014) for additional details.

Although the expressions above provide an exact solution for L_k, the complexity of computing this solution is likely intractable (as the problem is known to be #P-complete[1]), making it prohibitive for practical problem sizes related to P&R missions within DoD. To address this issue, from the analysis and results in Jung et al. (2008) and Jung et al. (2014), the key probability measure can be accurately estimated as

$$L_k = 1 - \frac{\mathbb{E}[n_k]}{v_k}, \tag{A.3}$$

$$\mathbb{E}[n_k] = \sum_{l \in \{n_k : \boldsymbol{n} \in S(\mathrm{C})\}} l\,\mathbb{P}[n_k = l] \approx \frac{\sum_{l \in \{n_k : \boldsymbol{n} \in S(\mathrm{C})\}} l \exp\left(q\left(\boldsymbol{x}^*(l,k)\right)\right)}{\sum_{l \in \{n_k : \boldsymbol{n} \in S(\mathrm{C})\}} \exp\left(q\left(\boldsymbol{x}^*(l,k)\right)\right)}, \tag{A.4}$$

$$q(\boldsymbol{x}) = \sum_{k \in K} x_k \log v_k + x_k - x_k \log x_k, \tag{A.5}$$

where $\mathbb{E}[\cdot]$ denotes the expectation operator, $\mathbb{P}[Y]$ denotes the probability of event Y, $\boldsymbol{x} := (x_k)$ is a continuous relaxation of $\boldsymbol{n}$ defined over a continuous relaxation $\bar{S}(\mathrm{C}) := \left\{\boldsymbol{x} \in \mathbb{R}_+^{|K|} : \boldsymbol{A}\boldsymbol{x} \leq \mathrm{C}\right\}$ of the space of possible deployments $S(\mathrm{C})$ and subsets $\bar{S}_{l,k}(\mathrm{C}) := \bar{S}(\mathrm{C}) \cap \{\boldsymbol{x} : x_k = l\}$ of this relaxation, $\mathbb{R}_+$ denotes

[1] #P-complete is a computational complexity class that defines a problem to be #P-complete if and only if it is in #P and every problem in #P can be reduced to it by a polynomial-time counting reduction, that is, a polynomial-time Turing reduction relating the cardinalities of solution sets (Valiant, 1979).

the set of non-negative reals, $x(l,k) \in \bar{S}_{l,k}(C)$, and $x^*(l,k)$ is the mode of the distribution associated with $\bar{S}_{l,k}(C)$. Corresponding results under general renewal arrival processes (υ_k, σ_k^2) can be found in Lu et al. (2007) and Cao et al. (2011).

Example: Time Inhomogeneous Markov Processes

Another example is based on the use of discrete-time stochastic processes to model the evolution of capabilities and readiness of personnel resources over time, given uncertainty around the time-varying supply-side dynamics with personnel acquiring skills, gaining experience, changing roles, and so on; some attrition of personnel; and new personnel being introduced. As a specific illustrative instance related to P&R, analogous to the workforce application studied in Cao et al. (2011), consider a time inhomogeneous Markov process of the evolution of personnel resources over a finite horizon of $T + 1$ periods in which resources are aggregated into groups based on combinations of attributes of interest (e.g., rank, skill, experience, proficiency). Let J denote the family of all such possible combinations of attributes of interest for personnel resources. The time inhomogeneous Markov process is then defined over the state space of all possible numbers of resources that can possess each combination of attributes composing J, namely

$$\Omega := \left\{ \boldsymbol{m} \in \mathbb{Z}_+^{|J|} : \boldsymbol{m} \leq \mathbf{M} \right\}$$

where $\boldsymbol{m} := (m_1, \ldots, m_{|J|})$ and $\mathbf{M} := (M_1, \ldots, M_{|J|})$, with $M_j \leq \infty$ as an upper bound on the maximum number of resources possible in group $j \in J$. Define $\boldsymbol{y}(0) := (y_1(0), \ldots, y_{|\Omega|}(0))$ to be the initial distribution and $\boldsymbol{K}(t)$ the Markov kernel for time $t \in \{0, \ldots, T\}$. The kernels $\boldsymbol{K}(\cdot)$ incorporate the probabilities governing all transitions between states in Ω from one period to the next (derived from the output of predictive analytics), including introduction of new personnel, attrition of existing personnel, acquisition of skills, promotion in rank, gain in experience, and so on. Then the distribution (probability measure) of this time-inhomogeneous Markov process at time t is given by

$$\boldsymbol{y}(t) = \boldsymbol{y}(0)\boldsymbol{K}(0)\boldsymbol{K}(1) \cdots \boldsymbol{K}(t-1), \tag{A.6}$$

where $y_k(t)$ represents the probability of being in state $k \in \Omega$ at time $t \in \{1, \ldots, T\}$ given the starting distribution $\boldsymbol{y}(0)$.

Due to the combinatorial explosion of the size of the state space Ω, an exact analysis is prohibitive for all but very small values of $|J|$ and M_j. To address this so-called curse of dimensionality, one can consider a type of fluid limit scaling of the time-inhomogeneous Markov process that instead records the expected number of resources in each group, thus rendering a single state for each group of attributes. For each state $j \in J$ and $t \in \{0, \ldots, T\}$, define $y_j(t)$ to be the expected number of resources in state j at time t, $\iota_j(t)$ the expected amount of new introductions into state j over $[t, t+1)$, and $\alpha_j(t)$ the expected amount of attrition from state j over $[t, t+1)$, with $\boldsymbol{y}(t) := (y_j(t))$, $\boldsymbol{\iota}(t) := (\iota_j(t))$, and $\boldsymbol{\alpha}(t) := (\alpha_j(t))$. Further define $p_{jj'}(t)$ to be the stationary probability of transitions from state j to state j' over $[t, t+1)$, where $\Sigma_{j' \in J}\, p_{jj'}(t) \leq 1$. When state j attrition is positive, this inequality is strict and $1 - \Sigma_{j' \in J}\, p_{jj'}(t)$ represents the stationary probability that a resource leaves state j through attrition, such that $\alpha_j(t) = y_j(t)(1 - \Sigma_{j' \in J}\, p_{jj'}(t))$. The evolutionary dynamics of this time-inhomogeneous Markov process then can be expressed as

$$y_j(t+1) =$$

$$y_j(t) + \iota_j(t) + \sum_{j' \neq j} y_{j'}(t) p_{j'j}(t) - y_j(t) \sum_{j' \neq j} p_{jj'}(t) - y_j(t)\left(1 - \sum_{j'} p_{jj'}(t)\right) =$$

$$\iota_j(t) + \sum_{j'} y_{j'}(t) p_{j'j}(t),$$

or, in matrix form (using row vector notation),

$$\boldsymbol{y}(t+1) = \boldsymbol{\iota}(t) + \boldsymbol{y}(t)\boldsymbol{P}(t),$$

where $\boldsymbol{P}(t) := [p_{jj'}(t)]$, $\forall j, j' \in J$, and $\forall t \in \{0, \ldots, T\}$. Additional algebra yields

$$\boldsymbol{y}(t+1) = \sum_{s=0}^{t} \boldsymbol{\iota}(s) \prod_{s'=s+1}^{t} \boldsymbol{P}(s') + \boldsymbol{y}(0) \prod_{s=0}^{t} \boldsymbol{P}(s). \qquad \text{(A.7)}$$

The experience reported in Cao et al. (2011) suggests the accuracy of this solution to be within a few percentage points in predicting the evolution of capabilities and readiness of personnel resources for large organizations with many thousands of resources over horizons of a year or so.

MATHEMATICAL OPTIMIZATION UNDER UNCERTAINTY

A more formal definition and description of mathematical optimization under uncertainty is presented, followed by a few examples of such methods involved in decision-making problems related to those of P&R.

In the case of one-time decisions over a given time horizon, a general formulation of a single-period decision-making optimization problem can be expressed as

$$\textit{minimize} \qquad f(\boldsymbol{x}, \boldsymbol{y}) \tag{B.1}$$

$$\textit{subject to} \qquad g_i(\boldsymbol{x}, \boldsymbol{y}, \boldsymbol{A}, b) \in \mathbb{S} \tag{B.2}$$

where $\boldsymbol{x} := (x_1, \ldots, x_n)$ is the vector of decision variables, $\boldsymbol{y} := (y_1, \ldots, y_m)$ is a vector of dependent variables, $f : \mathbb{R}^n \rightarrow \mathbb{R}$ is the objective functional, $g_i : \mathbb{R}^n \rightarrow \mathbb{R}$ is a constraint functional, $i = 1, \ldots, k$, $\boldsymbol{A}$ is a matrix, b is a scalar, and $\mathbb{S}$ is the constraint region. The objective functional $f(\cdot)$ and constraint functionals $g_i(\cdot)$ define the criteria for evaluating the best possible results, while the vector $\boldsymbol{y}$, the matrix $\boldsymbol{A}$, the scalar b, and the constraint region $\mathbb{S}$ are based on the stochastic models of the system of interest. The relationships among these components of the optimization formulation are critically important and often infused with subtleties and complex interactions. Of course, the objective in (B.1) can be equivalently formulated as a maximization problem with respect to $-f(\cdot)$.

When the variables in (B.1), (B.2) are deterministic (e.g., a point forecast of expected future demand), then the solution of the optimization problem (B.1), (B.2) falls within the domain of mathematical programming methods, the details of which depend upon the properties of the objective functional $f(\cdot)$ and constraint functionals $g_i(\cdot)$ (e.g., linear, convex, or nonlinear objective and constraint functionals) and the properties of the decision variables (e.g., integer or continuous decision variables). On the other hand, when the variables in (B.1), (B.2) are random variables or stochastic processes (e.g., a distributional forecast of future demand or a stochastic process of system dynamics), then the solution of the optimization problem (B.1), (B.2) falls within the domain of mathematical optimization under uncertainty methods, the details of which again depend upon the properties of the objective functional $f(\cdot)$ and constraint functionals $g_i(\cdot)$ and the properties of the decision variables.

In the case of adaptive decision-making policies for dynamic adjustments to decisions and actions throughout the time horizon as uncertainties are realized, a general formulation of a multiperiod decision-making optimization problem can be expressed as

$$\textit{minimize} \qquad \sum_{t=1}^{T} f_t(\boldsymbol{x}_t, \boldsymbol{y}_t) \tag{B.3}$$

$$\textit{subject to} \qquad g_{i,t}(\boldsymbol{x}_t, \boldsymbol{y}_t, \boldsymbol{A}_t, b_t) \in \mathbb{S}_t \tag{B.4}$$

where $\boldsymbol{x}_t := (x_{1,t}, \ldots, x_{n,t})$ is the vector of decision variables for time $t \in \{1, \ldots, T\}$, $\boldsymbol{y}_t := (y_{1,t}, \ldots, y_{n,t})$ is a vector of dependent variables for time t, $f_t : \mathbb{R}^n \rightarrow \mathbb{R}$ is the objective functional for time t, $g_{i,t} : \mathbb{R}^n \rightarrow \mathbb{R}$ is a constraint functional for time t, $i = 1, \ldots, K$, $\boldsymbol{A}_t$ is a matrix, b_t is a scalar, $\mathbb{S}_t$ is the constraint region for time t, and T is the time horizon. The objective and constraint functionals $f_t(\cdot)$ and $g_{i,t}(\cdot)$ define the criteria for evaluating the best possible results, while the vectors $\boldsymbol{y}_t$, the matrices $\boldsymbol{A}_t$, the scalars b_t, the constraint regions $\mathbb{S}_t$, and time horizon T are based on the stochastic models of the system of interest. The relationships among these components of the optimization formulation are critically important and often infused with subtleties and complex interactions. Of course, the objective in (B.3) can be equivalently formulated as a maximization problem with respect to $-f_t(\cdot)$.

When the variables in (B.3), (B.4) are deterministic (e.g., a point forecast of expected future demand over time), then the solution of the optimization over time problem (B.3), (B.4) falls within the domain of mathematical programming and deterministic dynamic programming and optimal control methods, the details of which depend upon the properties of the objective functionals $f_t(\cdot)$ and constraint functionals $g_{i,t}(\cdot)$ (e.g., linear, convex, or nonlinear objective and constraint functionals) and the properties of the decision variables (e.g., integer or continuous decision variables). On the other hand, when the variables in (B.3), (B.4) are random variables or stochastic processes (e.g., a distributional forecast of future demand over time or a stochastic process of system dynamics over time), then the solution of the optimization over time problem (B.3), (B.4) falls within the domain of mathematical optimization under uncertainty and stochastic dynamic programming and optimal control methods, the details of which again depend upon the properties of the objective functionals $f_t(\cdot)$ and constraint functionals $g_{i,t}(\cdot)$ and the properties of the decision variables. In addition, a filtration $\mathcal{F}_t$ is often included in the formulation to represent all historical information of a stochastic process up to time t (but not future information), where both the decision variables $\boldsymbol{x}_t$ and the system are adapted to the filtration $\mathcal{F}_t$.

To illustrate aspects of the points noted above and described in Chapter 4, a few examples of mathematical optimization under uncertainty involved in decision-making problems related to those of P&R are presented. The first example concerns trade-offs among skill capacities and readiness of resources given uncertainty around the demand for such resources, which relate to P&R missions associated with optimal mix, readiness, staffing, and resource allocation. This material is based on the works of Bhadra et al. (2007), Cao et al. (2011), Jung et al. (2008), Jung et al. (2014), and Lu et al. (2007), where additional technical details can be found. The second example concerns trade-offs among the evolution of capabilities and readiness of personnel resources over time given uncertainty around time-varying supply-side dynamics, which relate to P&R missions associated with recruiting, training, retention, optimal mix, readiness, staffing, career shaping, and resource allocation. This material is based on the work of Cao et al. (2011), where additional technical details can be found. The third example concerns trade-offs among the dynamic allocation of capacity for different types of resources in order to best serve uncertain demand, which relate to P&R missions associated with recruiting, training, retention, optimal mix, readiness, and resource allocation. This material is based on the work of Gao et al. (2013), where additional technical details can be found.

Example: Stochastic Optimization

One such example is based on optimization of stochastic loss networks to determine the best capacity and readiness of resources given uncertainty around the demand for such resources. As a specific illustrative instance related to P&R, analogous to the workforce application studied in Cao et al. (2011), consider a time horizon modeled as a stationary loss network consisting of multiple personnel resource types and multiple project types. The stochastic loss network is used to model the risk of not being able to satisfy project demand due to insufficient capacity of one or more resource types available at the time the project products need to be delivered. Define a linear base reward function with rate r_k for successfully delivering products of type k, a linear base cost function with rate c_i for resources of type i, and linear cost functions with rates c_i^ι, c_i^T, c_i^R for introducing, training, and retaining resources of type i, respectively.

The following formulation of a corresponding stochastic optimization problem can then be expressed as

$$\textit{maximize} \quad \sum_{k \in K} r_k (1 - L_k) v_k - \sum_{i \in I} \left(c_i C_i + c_i^\iota C_i^\iota + c_i^T C_i^T + c_i^R C_i^R \right) \qquad \text{(C.1)}$$

$$\textit{subject to} \quad \text{(A.3), (A.4), (A.5)} \qquad \text{(C.2)}$$

where the decision variables are the components of the capacity vector $C = (C_1, \ldots, C_I)$ and C_i^E, C_i^ι, C_i^T, C_i^R denote the amounts of existing, introduced, trained, and retained resources of type i with $C_i = C_i^E + C_i^\iota + C_i^T + C_i^R$ for all $i \in I$. Alternatively, (A.1) and (A.2) can be used for the constraints in (C.2) when the problem size is sufficiently small. In either case, a constraint can be added of the form $L_k \leq \beta_k$ for all k with $\beta_k \in (0,1)$, which will tend to push the optimal solution toward increasing the resource capacities C_i in order to lower the loss-risk probabilities L_k. A nonlinear program solver can be used to efficiently compute the optimal solution to any of these instances of the optimization problem (C.1), (C.2), possibly also exploiting the additional results in Jung et al. (2008) and Jung et al. (2014).

A time-varying, multiperiod version of the above stochastic optimization problem has also been studied. Refer to Bhadra et al. (2007) for the corresponding results and details.

Example: Stochastic Dynamic Program

Another example is based on optimization of discrete-time stochastic decision processes of the evolution of capabilities and readiness of personnel resources over time given uncertainty around the time-varying supply-side dynamics. One can start with the stochastic dynamic program formulated with respect to a time inhomogeneous Markov process whose dynamics are governed by (A.6) and where the decision variables would be based on changes in the Markov kernels $\boldsymbol{K}(t)$ that provide optimal evolution of capabilities and readiness of personnel resources. However, the combinatorial explosion of the size of the state space Ω makes the solution of such a stochastic dynamic program prohibitive for all but very small values of $|J|$ and M_j.

Alternatively, as an illustrative instance related to P&R analogous to the workforce application studied in Cao et al. (2011), consider the optimization of decisions or actions within a fluid limit scaling of a time inhomogeneous Markov process that models the evolutionary dynamics of personnel resources over a finite horizon of $T + 1$ periods according to equation (A.7). Define $C_j(t)$ to be the expected cost rate for state $j \in J$ over $[t, t + 1)$ and $\mathcal{R}_j(t)$ to be the expected reward rate for state $j \in J$ over $[t, t + 1)$, with $\boldsymbol{\mathcal{C}}(t) = (C_j(t))$ and $\boldsymbol{\mathcal{R}}(t) = (\mathcal{R}_j(t))$. The expected cumulative costs and rewards of resources over the time horizon are then given by

$$\sum_{t=1}^{T}\left[\boldsymbol{\mathcal{C}}(t)\cdot \boldsymbol{y}(t)\right], \quad \sum_{t=1}^{T}\left[\boldsymbol{\mathcal{R}}(t)\cdot\left(\boldsymbol{y}(t)\wedge \boldsymbol{D}(t)\right)\right],$$

respectively, under appropriate independence assumptions, where $\boldsymbol{D}(t) := (D_j(t))$ denotes the vector of demand for every state $j \in J$ over $[t, t + 1)$ and $\wedge$ denotes the component-wise minimum operator (i.e., $x \wedge y := \min\{x, y\}$); here the assumption is made that rewards are accrued up to the minimum of the amount of resources and the demand for such resources.

Equation (A.7) can be rewritten into the following discrete-time linear dynamical system recursion (using column vector notation):

$$\boldsymbol{y}(t + 1) = \boldsymbol{y}(t) + \boldsymbol{B}(t)\boldsymbol{u}(t), \tag{C.3}$$

where $\boldsymbol{B}(t)$ captures the sparsity patterns of $\boldsymbol{P}(t)$, and $\boldsymbol{u}(t)$ denotes the vector of decision variables for actions related to transitions between states (e.g., training and promotion) and for actions to introduce and retain personnel. Define $\mathcal{P}(t) := [\mathcal{R}(t) \cdot (y(t) \wedge \boldsymbol{D}(t))] - [\mathcal{C}(t) \cdot y(t)]$. Then the formulation of a corresponding stochastic optimal control problem can be expressed as

$$\begin{aligned} &\textit{minimize} \quad \sum_{t=1}^{T} -\mathcal{P}(t) + \boldsymbol{d}(t-1) \cdot \boldsymbol{u}(t-1), \qquad \text{(C.4)} \\ &\textit{subject to} \quad \text{(C.3)}, \end{aligned}$$

where the decision variables are $u(0), \ldots, u(T-1)$ and $\boldsymbol{d}(t)$ denotes the vector of expected decision cost rates for all states over $[t, t+1)$. Given the linear dynamical system in (C.3), the summation can be stacked in the objective function (C.4), and this optimization problem can be reformulated as a (piecewise) linear program in terms of a decision variable vector

$$\boldsymbol{z} := [y(1), \ldots, y(T), \boldsymbol{u}(0), \ldots, \boldsymbol{u}(T-1)]$$

and a weight vector

$$\boldsymbol{w} := [-\mathcal{P}(1), \ldots, -\mathcal{P}(T), \boldsymbol{d}(0), \ldots, \boldsymbol{d}(T-1)],$$

subject to the constraints (C.3) expressed in a corresponding matrix form. A linear program solver then can be used to efficiently compute an optimal solution vector z^*, which yields optimal state vectors $\boldsymbol{y}^*(1), \ldots, \boldsymbol{y}^*(T)$ and optimal decision vectors $\boldsymbol{u}^*(0), \ldots, \boldsymbol{u}^*(T-1)$.

Example: Stochastic Optimal Control

Another example is based on stochastic optimal control of the dynamic allocation of capacity for different types of resources in order to best serve uncertain demand so that expected net benefit is maximized over a given time horizon based on the rewards and costs associated with the different resource types. As a specific illustrative instance related to P&R sourcing problems, and a specific instance of the general class of dynamic resource allocation problems studied in Gao et al. (2013), consider the stochastic optimal control problem of satisfying demand (e.g., for sophisticated engineering projects) over time with two types of resources, namely a primary resource allocation option (e.g., chief engineer) that has the higher net benefit and a secondary resource allocation option (e.g., journeyman engineer) that has the lower net benefit. A control policy defines at every time $t \in \mathbb{R}$ the level of primary resource allocation, denoted by $P(t)$, and the level of secondary resource allocation, denoted by $S(t)$, that are used in combination to satisfy the uncertain demand, denoted by $D(t)$, where $S(t) = [D(t) - P(t)]^+$ with $x^+ := \max\{x, 0\}$. The demand process $D(t)$ is given by the linear diffusion model

$$dD(t) = bdt + \sigma dW(t),$$

where $b \in \mathbb{R}$ is the demand growth or decline rate, $\sigma > 0$ is the demand volatility or variability, and $W(t)$ is a one-dimensional standard Brownian motion, whose sample paths are nondifferentiable (Karatzas and Shreve, 1991). Define the reward and cost function associated with the primary resource capacity $P(t)$ at time t as $R_p(t) := r_p \times [P(t) \wedge D(t)]$ and $C_p(t) := c_p \times P(t)$, respectively, with reward and cost rates $r_p > 0$ and $c_p > 0$ capturing all per-unit rewards and costs for serving demand with primary resource capacity, $r_p > c_p$, and $\wedge$ denoting the minimum operator (i.e., $x \wedge y := \min\{x, y\}$). From a risk-hedging perspective, the risks associated with the primary resource allocation at time t concern lost reward opportunities whenever $P(t) < D(t)$ on one hand and concern incurred cost penalties whenever $P(t) > D(t)$ on the other hand. The corresponding reward and cost functions associated with the secondary resource capacity $S(t)$ at time t are defined as $R_s(t) := r_s \times [D(t) - P(t)]^+$ and $C_s(t) := c_s \times [D(t) - P(t)]^+$, respectively, with $r_s > 0$ and $c_s > 0$ analogous to r_p and c_p. From a risk-hedging perspective, the secondary resource allocation at time t is riskless in the sense that rewards and costs are both linear in the resource capacity actually used.

The decision process $P(t)$ is adapted to the filtration $\mathcal{F}_t$ generated by $\{D(s) : s \le t\}$. Any adjustments to the primary resource allocation capacity have associated costs, where $\mathcal{I}_p$ and $\mathcal{D}_p$ denote the per-unit cost for increasing and decreasing the allocation of primary resource capacity $P(t)$, respectively. Then the formulation of a corresponding stochastic optimal control problem can be expressed as

$$\text{maximize} \quad \mathbb{E}\int_0^\infty e^{-\alpha t}\left[N_p(t)+N_s(t)\right]dt - \mathbb{E}\int_0^\infty e^{-\alpha t}\left[\mathcal{I}_p \cdot I_{\{\dot{P}(t)>0\}}\right]dP(t)$$
$$- \mathbb{E}\int_0^\infty e^{-\alpha t}\left[\mathcal{D}_p \cdot I_{\{\dot{P}(t)<0\}}\right]d\left(-P(t)\right) \tag{D.1}$$

$$\text{subject to} \quad -\infty < \theta_l \le \dot{P}(t) \le \theta_u < \infty \tag{D.2}$$

$$dD(t) = bdt + \sigma dW(t), \tag{D.3}$$

where the decision variable is the derivative of $P(t)$ with respect to time, denoted by $\dot{P}(t)$, and where α is the discount factor, $N_p(t) := R_p(t) - C_p(t)$, $N_s(t) := R_s(t) - C_s(t)$, and $I_{\{A\}}$ denotes the indicator function returning 1 if A is true and 0 otherwise. The control variable $\dot{P}(t)$ is managed by the control policy at every time t subject to the lower-bound and upper-bound constraints in (D.2), which capture the inability to make unbounded adjustments in the primary resource capacity at any instant in time. Note that the second expectation in (D.1) causes a decrease with rate $\mathcal{I}_p$ in the value of the objective function whenever the control policy increases $P(t)$, and the third expectation in (D.1) causes a decrease with rate $\mathcal{D}_p$ in the value of the objective function whenever the control policy decreases $P(t)$.

Suppose the per-unit cost for decreasing the allocation of primary resource capacity is strictly less than the corresponding discounted overage cost and suppose the per-unit cost for increasing the allocation of primary resource capacity is strictly less than the corresponding discounted shortage cost. Then, as rigorously established by Gao et al. (2013), the solution to the stochastic optimal control problem (D.1), (D.2), (D.3) has the following simple form for governing the dynamic adjustments to $P(t)$ over time. Namely, there are two threshold values, L and U, with $L < U$ such that the optimal dynamic control policy seeks to maintain $X(t) = P(t) - D(t)$ within a risk-hedging interval $[L, U]$ at all times t, taking no action (i.e., making no change to $P(t)$) as long as $X(t) \in [L, U]$. Whenever $X(t)$ falls below L, the optimal dynamic control policy pushes toward the risk-hedging interval as fast as possible—that is, at rate θ_u, thus increasing the primary resource capacity $P(t)$. Similarly, whenever $X(t)$ exceeds U, the optimal dynamic control policy pushes toward the risk-hedging interval as fast as possible—that is, at rate θ_l, thus decreasing the primary resource capacity $P(t)$. Here, the optimal threshold values L and U are uniquely determined by two nonlinear equations that depend upon the parameters of the formulation. Refer to Gao et al. (2013) for these and related results as well as for additional technical details.

REFERENCES

Bhadra, S., Y. Lu, and M.S. Squillante. 2007. Optimal capacity planning in stochastic loss networks with time-varying workloads. Pp. 227-238 in *Proceedings of ACM SIGMETRICS Conference on Measurement and Modeling of Computer Systems*, San Diego, Calif., June 12-16.

Cao, H., J. Hu, C. Jiang, T. Kumar, T.-H. Li, Y. Liu, Y. Lu, S. Mahatma, A. Mojsilovic, M. Sharma, M.S. Squillante, and Y. Yu. 2011. OnTheMark: Integrated stochastic resource planning of human capital supply chains. *Interfaces* 41(5):414-435.

Gao, X., Y. Lu, M. Sharma, M.S. Squillante, and J.W. Bosman. 2013. Stochastic optimal control for a general class of dynamic resource allocation problems. Pp. 3-14 in *Proceedings of the IFIP W.G. 7.3 Performance Conference*, Vienna, Austria, September 24-26.

Jung, K., Y. Lu, D. Shah, M. Sharma, and M.S. Squillante. 2008. Revisiting stochastic loss networks: Structures and approximations. Pp. 407-418 in *Proceedings of ACM SIGMETRICS Conference on Measurement and Modeling of Computer Systems*, Annapolis, Md., June 2-6.

Jung, K., Y. Lu, D. Shah, M. Sharma, and M.S. Squillante. 2014. "Revisiting stochastic loss networks: Structures and algorithms." Preprint.

Karatzas, I., and S.E. Shreve. 1991. *Brownian Motion and Stochastic Calculus, second edition.* Springer-Verlag.

Lu, Y., A. Radovanovic, and M.S. Squillante. 2007. Optimal capacity planning in stochastic loss networks. *Performance Evaluation Review* 35(2):39-41.

Valiant, L.G. 1979. The complexity of computing the permanent. *Theoretical Computer Science* 8(2):189-201.